The First Decade of the Twentieth Century

The First Decade of the Twentieth Century

The Burgess Shale of Modern Technology

Gordon B. Greer

To the Lofferty
Best Regards
Gordon Greer

iUniverse, Inc.
New York Lincoln Shanghai

The First Decade of the Twentieth Century
The Burgess Shale of Modern Technology

iUniverse, Inc.

For information address:
iUniverse, Inc.
2021 Pine Lake Road, Suite 100
Lincoln, NE 68512
www.iuniverse.com

ISBN: 0-595-30725-6 (pbk)
ISBN: 0-595-66193-9 (cloth)

Printed in the United States of America

Contents

Acknowledgments

As may be quite obvious from the text of this book, I admire greatly the works of Professor Stephen Jay Gould on natural evolution, both for their contents and for his writing style. Who would believe that paleontology could be made so interesting and understandable?

The staff of the Information Technology Department at Bingham McCutchen did its usual excellent job of keeping my word processing system processing words. The staff of the publisher has been helpful throughout the publishing process. My wife, Nancy, has again been most patient and understanding during the writing and publication process, and my son, Bruce Greer, has again displayed his remarkable sense of history with a number of very apt comments on the text.

Any errors in the following material are mine alone.

Belmont, Massachusetts
February, 2004

Introduction

It has been reliably estimated that the age of the Earth is well over four billion years. While initially it was very hot, by about three and one half billion years ago the Earth had cooled and stabilized enough to begin to leave evidence of events that we can see and interpret today. These evidentiary materials were few and far between, but the time span covered is so large that there are still numerous points of reference although paleontologists always hope for more.

The paleontologists have a reasonably good feel for the order in which various sedimentary strata of the earth were created and, over time, converted to types of minerals. Thanks both to long-term radio-active decay rates and to the location of the strata, they know the ages of those strata. Once we go back in time beyond about 30,000 years from the present, the events of Earth history are played out in the rare anomalies appearing in rock or soil. The latter is most likely to be the source of early stone tools and/or bones of early members of the human race. The stone tools became quite sophisticated and well made beginning some 10,000 to 15,000 years ago (the Neolithic Era-new stone age and the Mesolithic Era-middle stone age) becoming much less so back to perhaps 500,000 or more years ago (the Paleolithic Era—old stone age) when the forms are such that it becomes difficult to tell whether a particular rock has been made into a tool by a person or was just an accidental shape. As we proceed further back in time, the fossil rock record of living organisms becomes the only means of determining what had occurred in the development of life on Earth.

While the fossils record the existence of single-celled life (algae) for a period of several billion years and bacteria may have preceded algae, multicellular organisms did not appear until much later. At sometime between 700 and 650 million years ago, some ribbon-like and other flat multicellular structures appeared in the fossil record (the Ediacara creatures). Those creatures seem to have been evolutionary failures, for nothing like them appears in the Cambrian or later strata.

The next burst of new types of life, and of much greater variety, occurred in the mid-Cambrian Period at about 530 million years ago, the Cambrian being the earliest (lowest) period in the Paleozoic (old life) Era. A small area of rocks in British Columbia discovered, fittingly in the first decade of the twentieth century and named the Burgess Shale, preserves the forms of perhaps two dozen animals not seen before the mid-Cambrian. Although four were identifiable with difficulty as probable predecessors of subsequent forms of invertebrate life and one (*Pikaia gracilens*) may possibly be the first chordate (animal with a spinal cord), most bore no resemblance to later creatures.[1] The Burgess Shale fossils were created in the shallow warm salt waters of the mid-Cambrian and were a portion of what has become known as the Cambrian explosion, when life forms of great diversity appeared suddenly (at least by geologic standards). But most of the Burgess forms and the similar but slightly earlier fossils recently discovered in China were small, fragile and seemed most unpromising to develop into the enormous varieties of present day life. Nonetheless, they did fill the waters, land and air with what we see around us today, by way such intermediate steps as dinosaurs and pterosaurs. Out of small beginnings...

---

1.   The late and much revered Stephen Jay Gould in his book on the Burgess Shale, *Wonderful Life* (1989), describes the fossils in the Burgess Shale in terms that even non-professionals in the field can readily understand. Professor Gould was extraordinary in his ability to do this in so many different areas.

1

Decades

This book is about the first decade of the twentieth century and its enormous impact initially on the technology and subsequently on almost all aspects of life in the later portions of that century. The first decade is usually viewed in social or geopolitical terms as the end of an era and the beginning of another very different and terrible one; even the *World Almanac* in its world chronology calls it the "Last Respite." That usual view will be discussed in the next chapter, although the thesis of this book is that five, largely embryonic, technologies from the first decade (automobiles, airplanes, nuclear reactions, HMS *Dreadnought* and radio) were really the most important aspects of the first decade. First, however, we must consider of what that decade consists. This is by no means as easy as one might think.

A decade is, of course, a ten-year period. The term is derived from the Latin word for ten-*decem*. The ten years should be contiguous. While a decade can begin in any year, it usually starts or ends with a year the last digit of which is 0. Some decades receive common nicknames-the Roaring Twenties; the Turbulent Sixties. These can be characterized by the fact that the penultimate digit is the same. So far, so good. But what years are contained in the first decade of the twentieth century? If there were a term for the penultimate digit 0, all could be made clear. We know, for example, that the Roaring Twenties began on January 1, 1920, and ended on December 31, 1929, because that period encompasses all of the years in that century which are "twenties." There is, unfortunately, no conventional collective term for zeros as there is for twenties. We must therefore designate the period by

1

description, "first decade of the twentieth century," and the matter becomes much more complicated. Did that decade begin on January 1, 1900, or on January 1, 1901? Public opinion would almost certainly favor 1900 as it did for 2000 as the millennium date. Public opinion would, however, be wrong as it was for the millennium date, and for the same reason; the first decade of A.D. (*Anno Domini*-Year of the Lord, or under the more recent and more politically correct system, C.E. (Common Era)) began with the year 1, at least after the calendar had been reconfigured, notwithstanding the fact that the *World Almanac* lists its decade segments from year 0 to year 9. There is good reason for the confusion.

In the sixth century, Pope St. John I was greatly troubled by the fact that the Christian church was reckoning by using a calendar inherited from the Roman Empire. That calendar counted years from the traditional date of the founding of the city of Rome (thus the date of Julius Caesar's death would have been, for the Romans, 709 A.U.C. (*Ab Urbe Condita*-From the Founding of the City) and not the date we now use of 44 B.C. (Before Christ, or again under the more recent and more politically correct system, B.C.E. (Before the Common Era)). The mythical date of the founding of Rome would, under our present calendar, be 753 B.C.). Not only the church but also most of the Mediterranean basin and northwest Europe used the same Roman-based calendar. There were other calendars in use around the world in this era. Lunar calendars were common in the Middle East, the Jewish one based on Biblical calculations and the soon to be adopted Islamic one based on the date of Mohammed's departure from Mecca, and some very sophisticated ones in the East and Mesoamerica, also usually at least partly lunar.

The Roman calendar, a combination of lunar and solar, had been corrected several times, most notably by Julius Caesar no less (the Julian calendar), because the Romans had previously miscalculated the length of the solar year resulting in serious slipping of the seasons into the wrong parts of the year. Apparently the Romans were better engineers than astronomers. Even Caesar's correction, or more fairly that of his Alexandrian consultant, did not get it quite right and the length of the year had to be corrected again under the auspices of Pope Gregory in the sixteenth century (the Gregorian calendar) to the system we use today. While Caesar could order the correction of the calendar and have the correction universally and instantaneously adopted within the Roman world, Pope Gregory did not have the power to require such a broad and rapid response to his changes. The Roman Catholic countries did accept fairly promptly but Protestants and Orthodox Catholics delayed so long that what started as a ten-day correction had become eleven days by the time the British acted in the mid-eighteenth century and thirteen days when the Russians acted in the early twentieth. Although it is now only history, the thought of operating in world with multiple slightly different calendars in regular use is troublesome. Of course we do something like this today but, at least in the West, the alternative calendars are largely liturgical and thus do not create too much confusion in everyday life.

Let us return to Pope St. John I (523 to 526 in the current calendar) and his problem with a calendar the defining date of which was the mythical founding of Rome. By this time the Roman Empire in the West was gone and the Pope felt that the birth date of Jesus was a more appropriate reference point than a heathen mythical one. This idea was in all likelihood not a new one; other Popes had doubtless felt the same. But so long as the Empire had stood in the Western Mediterranean, even though Christianity was widespread and officially endorsed after the conversion of the Emperor Constantine in 337, changing the historic base of the calendar was probably not feasible.

Now that the Empire in the West had fallen, there was an opportunity. The Pope designated a monk named Dionysius Exiguus or Dennis the Short to create a new calendar.[2] One might wonder how short he could have been in an era in which five foot three or four was con-

---

2.   A far more complete account of this effort can be found in Gould, *Questioning the Millennium* (1997).

sidered tall. Nevertheless Dennis researched at length the question of when Jesus had been born. There having been no Bureau of Vital Statistics operating in Bethlehem at the appropriate time, Dennis' task was very difficult and full of uncertainties. In fact, we should probably find it difficult today to obtain the precise date of an event five hundred years ago, particularly if it happened in a remote part of the world and was not considered very important at the time it occurred.[3] Dennis did what he could under the circumstances and eventually he arrived at his best judgment, although apparently influenced greatly, and erroneously, by a coincidental concurrence of the complicated calculation of the date of Easter Sunday. The calendar was shifted and, in addition to having new dates for everything, we now had the concept of B.C. and A.D., or in the new form, B.C.E. and C.E.

While little Dennis' calendar with its basing point as the assumed date of Jesus' birthday is followed at least in the West today, other basing points have been used on occasion even fairly recently. France, although a largely Catholic country, decided that its revolution was a more appropriate starting point and for some years used 1792-93 as year 1. The revolutionary years began in what had been mid-September. Occasionally one sees dates in the United States shown as "year X of the republic" but always in conjunction with a conventional date. After their revolution, the French tinkered a bit more with the calendar by renaming the months. That change, although also short-lived, was perhaps more logical. The conventional calendar then as now had four months named for their order of appearance in the year (September through December). As a matter of nomenclature those months should have been the seventh through the tenth months. The other eight months were and are named for rather outdated Roman gods (January through May, and possibly June) or demigods (July and August, for Julius and Augustus Caesar). The French substituted names for the

---

3.   *E.g.*, the birth date of Johannes Gutenberg, the inventor of Western printing from movable type and thus the father of modern printing, can be narrowed down only to the last decade of the fourteenth century.

months appropriate to the time of year (*e.g., Thermador*-heat; *Pluviose*-rain). Unfortunately this rational scheme was discarded during Napoleon's reign when the calendar was restored to Dennis' model.

On occasion, other basing points are used for specific purposes. One of the more common is the measurement of time in terms of years since the ascension of a ruler, as the fourth year in the reign of Henry VIII, or other memorable beginnings. A large university with which the author has had some contact dates many of its important events, including graduations, both by the conventional system as well as by the number of years since the founding of the university. With uncharacteristic deference, it gives the conventional system priority of place.

Unfortunately, little Dennis had not gotten it quite right in at least two respects. First, and quite embarrassingly, after the birth of Jesus Herod, King of Judea and a first-class villain of the New Testament, performed some acts described in the Bible. For example, he had summoned before him the Magi who had been following the star toward Jesus' birthplace and he had ordered the execution of first-born sons in hope of killing the newly-born King of the Jews.[4] In addition, Joseph, Mary and Jesus were reported in the Bible to have been living in Egypt when Herod died.[5] Subsequent to Dennis' calculations it became clear that Herod had died in what had then become known as 4 B.C. so Jesus had to have been born at least four years before he had been born. Either Dennis or the Bible was in error. It will not come as a great surprise that the church decided the error was Dennis' but his calculations were not adjusted. Secondly, and of more importance at least to the present issue, Dennis' calendar years ran 2 B.C.,1 B.C.,1 A.D.,2 A.D. etc. There was no year 0, or possibly years 0, one in the A.D. sequence and one in the B.C. sequence. That second complication, while important, can be ignored here because we are only concerned with the A.D. years. But in a number sequence the digits run 0, 1, 2 etc. Consider

4.   *Bible, New Testament*, King James Version, Gospel According to Matthew, ch. 2, v. 7-17.
5.   *Ibid.*, ch. 2, v. 19-20.

that one is not thought to have been born on one's first birthday although the literal wording would suggest that such was the case. Logically one's birthday is the day of one's birth; one's birth date might commemorate the particular date and month. But what we call birthdays are really anniversaries of the birthday or celebrations of the birth date. By common convention one's first birthday comes after a year (the year 0) of one's life. Nevertheless, the first decade in the A.D. portion of little Dennis' invention must start on January 1, 1 A.D., because all dates preceding it are B.C. Thus the first decade of A.D. must end on December 31, 10 A.D., in order to consist of ten years. Unless we propose to create gaps or overlaps in the sequence, all sequential decades, plus centuries and millennia, must also begin in years the last digit of which is 1 and end with years ending in 0. Therefore the first decade of the twentieth century runs from January 1, 1901, to December 31, 1910.

We should not be too critical of little Dennis. He was doing all of his calculations in Roman numerals. Apart from the obvious difficulties of doing mathematical work of even the most basic kind in Roman numerals, the concept of the numeral 0 or anything similar was unknown in Western Europe or almost anywhere else at the time of Dennis' work. There has been a great deal of scholarly research on the concept of zero, or more accurately on the evolution of the several concepts embodied in our zero. For those brave enough to want to investigate this matter further, there is nothing (sorry) better than Robert Kaplan's recent work, *The Nothing That Is, A Natural History of Zero,* Oxford University Press (2000). For purposes of this book, however, it is enough to say that Dennis had not heard of zero nor had anyone with whom he could have had contact known about the concept.

The idea of zero along with what we call Arabic numerals, although they may well trace back to India, arrived in Catholic Europe around the year 1,000 via the Moors in Spain and a Pope, Sylvester II, who had lived in Spain. While the transition from Roman numerals to Arabic numerals had enormous benefits when completed, the process must have been very uncomfortable for those attempting it. Making the transition more difficult to follow, the Arabic system uses 0 to represent the numeral we call five; the zero concept is represented by a dot.

The transition from Roman numerals to Arabic (or to some scholars, to Hindu-Arabic) numerals by Christian Europe was by no means a simple substitution of one counting system for another. If one compares the present standard Arabic numerals to the standard Western ones, only 1 and 9 are identical. The numerals 2, 3 and 7 in Arabic can, with some change in orientation which has an historical basis, be identified with their Western counterparts. But 4, 5, 6, 8 and 0 do not seem to bear any family resemblance. That should make little difference to us because the real gift of the Moors was not so much their system of numerals but rather the application of that system to a method of counting using a base and thus using numbers the value of which was determined by position rather than by aggregation. They used base 10 which we adopted. Base 10 was not critical although possibly helpful to those who wished to count on their fingers. Any base would have served well in the long run as a replacement for Roman numerals although numbers between 8 and 20 would probably have served best for most calculations. In fact today for some purposes we use other bases, generally 2, 8 and 16.

The disparity between today's Arabic and Western numerals does not entirely reflect an historical difference introduced by Europeans for some reason, perhaps chauvinism, a thousand years ago but rather reflects in large measure an evolutionary process of both systems over the past thousand years. In fact, the Arabic system seems to have changed more in that period. In the first example of Hindu-Arabic numerals in Europe (dated 976), the numerals 1, 6, 7, 8, and 9 are identical to our present forms and 2 and 3 are close,[6] or a better correspondence to present Western numerals than the present Arabic system as described above.

---

6.   Ifrah, *From One To Zero*, Penguin Books English Translation, p. 477 (1987) (the *Codex Vigilianus*).

In due course the new numeral system including the revolutionary concept of zero had percolated through the rest of Europe, at least among serious mathematicians. But viewing the situation from our time when mathematics is taught universally to grade school students at levels that, even after the adoption and widespread use of the Arabic system, could not have been understood by most university graduates a few hundred years ago, it is difficult to appreciate the level of mathematical ignorance in the latter part of the Middle Ages. There is a story about a German merchant in the fifteenth century who inquired of a university professor where he should send his son for mathematical training. The professor replied that if addition and subtraction were all that were desired any university would suffice, but if multiplication and division were needed those functions had been greatly developed in Italy and his son should go there to study.[7] So as not to defame unduly the intellect of the Late Middle Age mathematicians, it should be noted that their techniques for those operations were very much more cumbersome that the ones in use today.

Notwithstanding all of the foregoing and to avoid rehashing the millennium debate for purposes of this book the first decade of the twentieth century (the "First Decade") will be considered to have begun on January 1, 1900, and to have ended on December 31, 1910. This will encompass all of the years contained in either definition of the period. Those whose arithmetic skills have not totally atrophied in this calculator-dependent age will recognize immediately that this is an eleven-year period and that I have done violence to the term "decade." So be it.

---

7.   Ifrah, *op. cit. supra*, p. 432.

2

Geopolitics

As previously stated, this book is about the technology of the First Decade although most historians view the First Decade largely in geopolitical terms. There is ample justification for their approach. In discussing the technology of the First Decade it is necessary to have some appreciation of the underlying geopolitics not so much for their impact on the First Decade but rather for their impact upon the development and use of that technology in subsequent decades—war has the effect of putting great pressure on technical progress and accelerating it many-fold. In that sense a review of the geopolitical situation, in Europe particularly, is in order.

Relationships between countries seemed to be changing at a more rapid rate during the nineteenth century. Speed of communication, the telegraph and the telephone, and speed of travel, the railroad and the steamship, certainly would accelerate the building of crises but should also have hastened resolution of them. It may be, of course, that we tend to see the events and relationships nearest to our time in the most detail and that the further back in time one looks the thinner the record appears to us, but perhaps it does not seem the same to those living at the time. Nonetheless, the apparent relative stability of some ancient empires is impressive, particularly by modern standards; Egypt, Persia, China, Japan, India, Sumer, Babylon and in some ways the most stable, Rome. They each persisted at least for hundreds of years, sometimes for thousands, as important entities albeit with changes, sometimes dramatic, in their territories and/or structures. In recent times the rate of change, particularly in European countries, seems to have accelerated even after including those changes occurring in earlier periods because of the episodic influx of various peoples from the East into the European continent prior to the sixteenth century.

Let us consider for a moment the changing status of a number of countries from the American Revolution to the beginning of World War I.

The British Empire-While England appeared as an important player on the world's stage and the most powerful Protestant nation in Europe as a result of its defeat of the Spanish Armada in 1588, it continued to spar for many years with France, Scotland, Ireland and The Netherlands, finally uniting the British Isles, more or less, and making its major error in losing the United States. The remainder of its vast colonial empire, however, remained intact. The British continued their traditional wars with France into the mid-eighteenth century and, after a brief respite in the late 1700s, came back into the fray against the new republican government as part, at times, of a coalition of European powers which coalition also varied at times. The long series of wars against the republican government and the Napoleonic empire continued until 1815 when the previous British, Austrian, Prussian and Russian allies were reassembled on short notice as only Britain and Prussia to combat Napoleon on his return from Elba after his first exile. The resulting Battle of Waterloo marked the emergence of Great Britain as the paramount power not only in Europe but in the whole world. In part because of its overwhelming power, it did not have to fight major wars between Waterloo and World War I, although there were a number of minor (except to those involved) wars; Crimea, Sudan, the Indian mutiny, various fights in northwest India and Afghanistan and some small actions in China, for instance. By the end of the nineteenth century Great Britain controlled a quarter of the world's land area, the Royal Navy was by far the largest and most powerful navy in the world and, not coincidentally, Britain controlled most of the world's shipping trade.

France-After Waterloo, France avoided initiating European military adventures, concentrating instead on her colonial empire in Africa, Southeast Asia and the Pacific when not distracted by internal troubles. She did participate with Britain and Turkey in the Crimean War against Russia which the allies were deemed to have won, although not by much. In 1870, France lost the Franco-Prussian War ignominiously and thereafter devoted most of her military resources to try to prevent further incursions by Germany (*nee* Prussia).

Germany-It came into being in 1871 as a result of von Bismarck's efforts in mid-century to assemble all of the Germanic peoples into one nation. The labor pains of this creation involved, apart from the serious and protracted political negotiations, three wars in quick succession. In 1864, Denmark was attacked and easily defeated, losing some territory deemed strategic to Prussia; in 1866 Austria was attacked and defeated; and in 1870-71 France was soundly trounced with Paris being captured. As a result some previously French territory on the border of what was becoming Germany now became German. Bismarck's creation began to take shape as the Second Reich with a Kaiser (William I) at its head. Although Bismarck had done an impressive job in creating the Second Reich, he was dismissed early in the reign of its second and last Kaiser, William II. The First Reich had been the entity created by Charlemagne and Pope Leo in 800, called most unrealistically the Holy Roman Empire, and the Third Reich we know all too well. Because the Second Reich was late in bringing its new empire to the overseas colony game, not much was left then unpicked and even many of the former colonies of other countries by then had become independent. Thus, despite its high aspirations, the new nation was left with a few not very desirable areas in southern and western Africa and a number of minor Pacific islands. While these islands did not seem to do much for the Second Reich, after World War I they were awarded to Japan, giving the United States considerable difficulty in World War II when it was obligated to capture or neutralize them. But Germany's strength was its scientific and manufacturing prowess as well as the historical excellence of the Prussian army. That excellence had been, of course, well displayed again to the world in the Danish, Austrian and French wars. The dramatic German successes in those wars not only disturbed Germany's other neighbors greatly but also encouraged German strengthening of its industrial power and its desire for more territory. It also seemed to have generated a wide-spread feeling of great patriotism. A friend of the author, in addressing the senior staff of an American company recently acquired by his German company, was describing the history of that German company. In noting that it had been established in 1871, he said that was a year in which many prominent German businesses had been created. He ascribed that year's business popularity and general euphoria to the fact that 1871 was the last time Germany had won a war. In aid of its various aims, one aspect of German militarism marched off in a new direction-toward the high seas. Initially the new German navy was such a minor factor in the overall military forces of the Second Reich that it was headed by a general. In short order, however, von Tirpitz, a legitimate naval officer, took over command and the rapid expansion of the Kriegsmarine began, leading to its eventual confrontations with the Royal Navy. Chapter 6 deals with the technological aspects of that rivalry.

Russia-Its actions during the winter of 1812-13 started the final decline of Napoleon. Russia, as always, was greatly assisted by the big three Russian military assets: long, cold winters; large, tough armies; and huge buffer land areas. For the rest of the nineteenth century Russia changed little, thus falling further behind Western Europe. Its scientific capacity was spotty although with a few brilliant flashes, its industrial base was very poor, especially given the plethora of natural resources available, and its politics were quite unstable, particularly for an absolute monarchy. Russia's principal wars between 1815 and 1900 were wars with Turkey, including the Crimean but its loss there did not seem to have much of an effect on the country. Russia at the end of the century was very much like Russia at the beginning of the century, although it had made considerable progress in taming parts of the Siberian wilderness and had begun work on the transsiberian railroad (finally completed to the Pacific in 1916). Russia's situation was about to change in ways that influenced greatly the remainder of the century.

Turkey-Not for nothing was the Ottoman Empire called the "sick man of Europe." During the nineteenth century that formerly great Empire was gradually disintegrating by losing its Balkan and Greek provinces. While parts of its army were quite good, due to its absence of expansionistic activities Turkey was not perceived by the West as being of great military importance except for its strategic value in controlling access to the Black Sea.

Austria-Hungary-This strange, at least for the late nineteenth century, polyglot empire was made up of Slavs of various kinds, Germans and Magyars. It was also politically unstable although ruled by the Hapsburgs as heirs, once removed by virtue of Napoleon I's usurpation, to one of the oldest crowns extant, that of the Holy Roman Empire. The country was strange in another way in that, notwithstanding the "Roman" in the ancient title, there were not included in it any Romans or even a significant number of Italians. Austria-Hungary had some superb military units but ethnic tensions were severe and the government was generally quite ineffective. In its last major war it had been defeated by Prussia.

United States-It was not a factor in world politics in the nineteenth century, having generally followed George Washington's advice to avoid foreign entanglements but his successors did make exceptions for Latin America beginning with the rather high-handed and unilateral assertion of the Monroe Doctrine and for Japan in a very mild fashion in mid-nineteenth century. That century in the United States was generally devoted to overwhelming internal matters, the Civil War and related events and the settling of the West. Several small external wars were fought. The War of 1812 with England was lost but with very little cost to the United States, and the Mexican War was won to the great territorial benefit of the United States. After the Civil War, the army was reduced to regimental-sized units to fight the Western Indians and the navy, which briefly had been technologically the best in the world, seemed to its great detriment to have been frozen in time in 1865. At the end of the century Spain and the United States went to war, resulting in an imperialistic surge (the Manifest Destiny concept) in the United States and the effect on Spain described in the following paragraph. Pushed to some extent by Mahan's views on sea power and Theodore Roosevelt's ambition, the United States had finally embarked on a naval modernization shortly before the war with Spain, to America's great advantage in that war. Its naval successes there encouraged the general improvement of the navy which had found its overseas responsibilities much increased by the acquisition of the previously Spanish-held territories as well as by the annexation of the Hawaiian Islands slightly earlier. The result was an effort by the United States to create a navy on a par with the major European powers. It did not succeed in approaching that of either Germany or Great Britain by the beginning of World War I, but it did have a respectable sized and very modern navy by 1914.

Spain-Unfortunately it had suffered a disastrous nineteenth century. From being the most prosperous and powerful nation in Europe in the sixteenth century it saw its fortunes decline, beginning with the defeat of the Armada by the English fleet with a big assist from the weather. That decline continued in the nineteenth century with the loss of most of its colonies in the Western Hemisphere in the first part of the century. The loss was completed in 1898 as a result of the Spanish-American War which allowed Cuba to become independent, the Philippine Islands to become wards of the United States and Puerto Rico and Guam to become American possessions. Thereafter, Spain was not much of a factor, political or military, in Europe.

Italy-The country became unified in the mid-1800s after centuries of fragmentation and occupation by various foreign powers. The timing of this rather paralleled the unification of the Second Reich. The Germans, however, progressed further and faster in pursuit of an integrated population and economy. The Italians had a similar problem to the Germans in being late in the search for colonies, but the Italians stayed in central and northern Africa and the Mediterranean, perhaps because it was more convenient, perhaps because of easier logistics or perhaps because it harked back to the glory days of the Roman Empire. Further suggesting the last was the later name of the ruling political party, the Fascists, taken from the Latin *fasces*, a symbol of governmental authority in ancient Rome. Italy's continued uncertainty as to where its best interests lay was further reflected in its decisions to fight on both sides in both World Wars.

Japan-In prying Japan out of her feudal and isolationist past in mid-nineteenth century, the United States managed unintentionally to initiate the creation of a powerful and tenacious enemy a hundred years later. At the turn of the century, however, Japan was just beginning to appear on the world's stage or perhaps then only in the wings. While no particular external events had given the Western world reason to expect major events to happen in the Japanese sphere (and if it did it might well not have been much interested), the Japanese had been learning a great deal from the West. They understandably were much concerned with naval matters and, with help from Great Britain, had begun to build a modern fleet. The Russo-Japanese War in 1904 would shortly show how well the Japanese had progressed in naval technology and tactics as well as giving some hint as to their longer-term ambitions.

Barbara Tuchman in her wonderful description of the funeral of King Edward VII of England in 1910[8] points out both the close inter-relationships of the ruling families of the European monarchies and their family rivalries but, knowing the way the next few years would play out, she appropriately treats the event as the "last hurrah" of the nineteenth century's relative political stability.

Also with the benefit of hindsight we can see that the few ripples of waves in the apparent tranquility since 1815, the revolutions of 1848, the Prussian Wars of 1864-1870 and the instability in Russia, were

---

8.   Tuchman, *Guns of August*, ch. 1 (1962).

warning signs for the structure. From the list above one can readily see that power, as always, was waxing and waning among the European powers but perhaps more rapidly than had been usual in the past. Yet, with the exception of some small territorial changes, on the surface nothing much had been altered since the Congress of Vienna had restructured Europe after the Napoleonic Wars. But since the relative strengths of European nations had changed, the strains were there and something was liable to crack. It was not unlike the shifting of the earth's crust. Once the shifting starts, strains are set up and stresses are created which must be relieved at some point. To date it cannot be determined with any accuracy when this will happen but happen it will and an earthquake will result, removing the stresses temporarily until further movement reestablishes them. So too European geopolitics but, unlike nature, politicians and rulers perceive themselves to have a very limited time frame, causing events to happen considerably more rapidly than nature needs to arrange.

The earthquake analogy can be carried a bit further. While small quakes can be considered a form of warning of larger quakes to come in the area, the fact is that the small quakes release some of the energy that would otherwise be likely to appear in the later, larger earthquake. The major European powers spent the thirty years or so before the beginning of World War I creating and altering a series of alliances, changing their terms and composition to adjust to various crises and perceptions. By putting out or sweeping away brush fire problems, some small wars were avoided temporarily but the issues behind those possible wars contributed to the larger issues which seemed to spur the combatants on to war later. Most of these involved German expansionism as did the grave potential of the accelerating domino effect of the Triple Alliance and the Triple Entente.

If one accepts the proposition that the stresses in European geopolitics would have to be relieved at some point, the issue becomes when and, most importantly, how. Again, Chapter 6 deals with one unrealized possibility.

3

Automobiles

It seems strange that the automobile was not created sooner than it was. After all, "automobiles" in the literal sense of vehicles which could be powered and directed on internal machinery and fuel, had been around in one form or another for all of the nineteenth century and in fact they probably could be thought to have begun in the late eighteenth with John Fitch's awkward and inefficient steamboat in the United States followed by William Symington's steamboat in Scotland.[9] Although Fitch's steamboat worked, because he attempted to duplicate the action of a paddling a canoe, steam was inefficiently harnessed to move oars. A few years later Robert Fulton invented a steam-driven boat which was propelled by two paddle wheels, one on each beam. This method of propulsion resembled that of Symington's and was obviously far more efficient than Fitch's, made even more so over time when it became possible to reverse one or both paddle wheels. This enabled the vessel to stop or change direction very rapidly, much more than with rudder control alone.

Because fuel and boiler water had to be carried by steamships (clean fresh water was required as salt, brackish or muddy water is corrosive and/or sediment leaving), the range of early steam vessels was quite limited, particularly in salt water. As a consequence, one of the major

---

9. There are reports of earlier efforts, one of which describes a sixteenth century Basque proposal to King Carlos I of Spain to develop a ship driven by a paddle wheel powered by vapor from boiling water. Kurlansky, *The Basque History of the World*, p. 56 (1999). His Majesty declined to participate, perhaps thereby losing a greater place in history.

early uses of steam vessels was as harbor tugboats. This is not as unimportant a job as it might seem to present day sailors used to seeing yachtsmen maneuvering their fore-and-aft rigged small boats into and out of dockage without use of their auxiliary engines. Even large modern steam or diesel powered vessels so commonly use tugs for docking that it is cause for comment when one docks under its own power, perhaps because of the unavailability of a tugboat. But none of this compares to the problem of moving large square-rigged vessels, both naval and merchant, in and out of dockage space. Just waiting for the right tidal conditions for an attempt at docking could take many hours; square-rigged sail handling was exceptionally labor intensive and very slow reacting as compared to fore-and-aft rigging. It was not just to prevent crews from deserting that many large, particularly naval, vessels anchored out in the harbor, coming ashore and being provisioned by small boats rather than directly from piers. Anchoring as opposed to tying up to a pier also made getting under way much easier. If the anchor were not fouled ships could depart pretty much on their chosen point of sailing rather than having to wait for an appropriate wind given the position of the pier. Even from an anchorage a large fleet under sail could require hours to take their leave of a harbor. The Battle of Trafalgar was a good example; the French and Spanish fleets took more than a day to clear Cadiz harbor and Cadiz Bay, giving the Royal Navy ample time to plan and maneuver.

For most of the nineteenth century steam slowly supplanted sails for both naval and merchant use, but it was not a smooth transition. The much heralded transatlantic voyage of the United States "steamship" *Savannah* in 1819 was conducted mostly under the sail rig that she also carried. A large part of the problem was that she could not carry enough fuel (and water) to make the trip entirely under power. But eventually even the Royal Navy discovered, during the Crimean War, that its unpopular steam frigates were far more useful than its traditional three-masted sail-powered line-of-battle ships. By the time of the American Civil War almost all major sea-going naval vessels on both sides were both steam and sail powered. Obvious exceptions were the USS *Monitor* and the CSS *Virginia* (*nee* USS *Merrimack*; yes, there is a final "k") and their relations. Gradually, however, the importance of the alternate sail rig diminished, hurried along by the development of armored iron and steel ships beginning with the French *Gloire* in 1859 and the HMS *Warrior* in 1860. These ships, while equipped with auxiliary masts and sails, were even harder to move under sail. Pure sailing vessels did survive for many years but only for use in certain parts of the merchant trades, tea, lumber and coal for example, until World War I.

Early in the transitional period from sail to steam a debate erupted about the relative merits of screw propellers versus paddle wheels as the more efficient method of applying power to move the vessel. Intuitively large side paddle wheels seemed more effective than screw propellers, but the propeller had its advocates. There were serious concerns about the vulnerability of paddle wheels in combat but screws were perhaps more prone to damage by grounding and were certainly harder to repair under way. Finally the Royal Navy decided on an experiment. Two similar ships were built, one equipped with side paddle wheels the other with a screw propeller. They were connected stern to stern and run up to full power. The screw propeller powered ship dragged the paddle wheeler backwards at about 3 knots, settling the matter at least so far as the Admiralty was concerned.

There was still much work to be done to create the most efficient propeller design, and side paddle wheels survived long enough to give rise to one of the more unusual maritime terms. In sailing ship days, ships were commanded from the rear of the ship, often on a raised quarterdeck right aft. From there the captain could see the effect of the wind on the sails and could easily give orders to the helmsman who was also positioned there both to view the sails as well so as to have the shortest connection between the wheel and the rudder located just beneath the stern. In emergencies if the linkage to the tiller had been broken, the captain could shout orders down below to manual operators of the rudder. Because of the position of the sails, forward visibility from the quarterdeck was not terribly good for either the captain or the helmsman as anyone who has tried to steer a small sailboat will know. With the advent of paddle wheels, viewing the sails, even if they had been set, became much less a concern and with higher speeds better forward visibility became relatively more important. At about the same time linkage between the wheel and the rudder was made more reliable by metal connections rather than by ropes and pulleys. Accordingly, a better site for the control of the ship was established on a raised platform connecting the tops of the two paddle wheel covers; a bridge over the deck below. The abolition of the paddle wheels did not cause the name of the raised forward control center to be changed even though it was no longer on a "bridge" but rather was incorporated directly into the superstructure of the vessel, and the bridge it is called to this day.

While paddle wheels declined for oceangoing vessels, they were retained on some inland waterways for shallow water use and easy repair purposes. Most are familiar with the big Mississippi River sternwheelers from the latter part of the nineteenth century but not so many realize that stern wheeled river tugs were common through World War II.

As steam engines became more widely used at sea, they became more reliable and smaller models became available for boats as small as launches. Whatever the size of the vessel or the engine, until almost the end of the century those engines were all of reciprocating design. What was meant by that is that the engine had pistons in cylinders like an automobile engine except that high pressure steam was introduced into the cylinders instead of a fuel-air mixture to move the pistons. The back-and-forth or up-and-down motion of the pistons then had to be converted to rotary motion by mechanical means in order to drive a propeller.

Thus it is clear that the concept of automotive devices of varying sizes and complexity for use on water was developed and progressively improved throughout the nineteenth century, although the transition from paddle wheels to propellers may indicate some deviation away from a method of locomotion that could have been easily adapted to overland travel.

Starting a little later in the century than the initiation of steam-powered vessels, another category of automotive devices, later known as locomotives, evolved. Beginning with George Stephenson's first practical machine in England in 1829 and spreading to the United States the next year, the railroad locomotive was born. While the first versions were rather primitive, the locomotives improved rapidly as did the rail network and the rolling stock in many countries, including the United States. The result was that, by the end of the United States Civil War, steam-engine railroads burning either coal or wood were available all over the country east of the Mississippi River and a few years later all across the country. Models of an automotive device for use on land were thus also widely available, and even if fuel for a steam automobile would still have been a problem the railroads had solved the water supply issue by building tanks along the railroad which fed water into the steam engines' boilers by gravity; the same could have been done by roadside facilities for automobiles.

Given that examples of automotive use on water and on rails were commonplace by mid-century, why was development of the modern automobile (in the sense now used) not initiated until the latter part of the century in Germany and the United States or improved over a very primitive French effort in the late 1700s? The French attempt looked rather like a form of farm wagon with a steam boiler on the front and was clearly an evolutionary dead end; the German was Benz's first car, powered by an internal combustion engine, and the American were those of Selden and Duryea similarly powered. Why did it take so long since the technology, at least of the steam variety, was well known? It was not that no one thought of this application. For example, as early as the 1820s a famous Scottish engineer named Thomas Telford favored road locomotion by steam over steam-powered railroads[10] but

---

10.   Herman, *How the Scots Invented the Modern World,* p. 336 (2001).

his view did not prevail, and William Symington had worked on steam power for "land carriages" before turning to steamboats.

Probably most of the delay was based upon two general categories of problems. The first, the weight of the machinery of a steam engine, plus the weight of sufficient water to travel a practicable distance (steam engines in those days did not recover the water from used steam), plus a reasonable amount of fuel for the boiler was, in the aggregate, a large amount to move up a grade on a road. On water there were no grades and on railroads the grades were gentle slopes which could be made with whatever pitch was necessary to permit the equipment to function, usually not much more than about 1% unless the railroad was of the cog-drive type, an inefficient system used only for steep mountain grades. On land the grades of roads were very steep and the power and traction necessary to deal with them was considerable. Second, the roads of the nineteenth century were virtually all dirt and very rough dirt at that. This not only required more power to move on and therefore more weight, in turn requiring more power, etc., but also a rough road surface could be very damaging to the machinery moving over it. Whatever the reasons, an automotive device for movement on land over ordinary roads was very late in coming. It was probably not a coincidence that a number of early automobiles were designed shortly after the four-cycle gasoline engine was invented in 1876. That engine, lighter than steam although more complicated and heavier when compared to a two-cycle gasoline engine of the same power, was efficient as well as having a much cleaner exhaust.

As the automobile was developing in the 1890s, many different forms of propulsion were considered by people who could only be described then as experimenters. After the rapid elimination of concepts which were going nowhere, such as compressed gases, springs and gravity, the preferred remaining forms were electric motors, steam via reciprocating engines and internal combustion engines using either spark or compression ignition (gasoline or diesel), also in reciprocating type engines. During this stage there was no clear winner in the propulsion sweepstakes.

Steam had the advantage of a huge amount of practical experience and an absence of any necessary speed-reducing gearing to transfer power from the engine to the wheels. Simply adjusting a valve to let more or less steam into the cylinder(s) of the engine provided whatever direct power was needed by simple mechanical linkage to the drive wheel(s). The cylinders plus a crankshaft to transfer reciprocating motion of the pistons to rotary motion were all that was necessary to convert steam pressure power to the wheels, and it would be fair to say that there was not much of a speed limit imposed by the engine.[11] On the other hand, the boiler had to be quite lightweight compared to ship or locomotive boilers but it carried a fair amount of pressurized steam and thus had considerable potential for explosion, particularly in an accident. Furthermore, the steam car had a significant delay in starting because the boiler had to be heated enough to generate steam before the car could move. One might have thought such a delay would be easily borne given the power which was available after the delay and the relative simplicity of the engine, but the late twentieth century experience of Mercedes with its diesel passenger cars was instructive in this regard. Mercedes had, by the mid-thirties, developed eminently practical diesel-powered passenger automobiles. They were conventional sedans of the same size and shape as some gasoline-powered cars of the same manufacturer and were produced in relatively small numbers throughout the century. The diesel had the advantages of long engine life, cheaper fuel and high torque ratings, but these were outweighed by the disadvantages of less power from the size and weight of engine (because in part the compression ratios were approximately three times those of similar gasoline engines, requiring much stronger engine

---

11. A steam-powered automobile, the Stanley Steamer, set a speed record of 127 miles per hour in 1906 using only a 30 horsepower engine. Clymer, *Early American Automobiles*, p. 20 (1950). With the steering and suspensions available in those days, it would have taken a very brave man to drive at that speed.

parts), an exhaust gas that looked and smelled worse than a gasoline car (but which was not as poisonous as were the carbon monoxide-laced gasoline exhausts), a louder engine noise and, perhaps most seriously, a delay in the starting time of the engine. This in spite of the fact that the delay was a maximum of about two minutes, far less than that of steam engines of that era.[12] Although steam cars needed fuel and water at frequent intervals, refueling and watering were quick and easy and restarting, while slow, was also easy.

Electricity had much to recommend it. It was quiet, odorless, efficient, easy to start, allowed easy engine (motor) placement and required nothing more than a simple mechanical link between the engine and the wheels to transfer the power where it was needed. It could also supply high torque at low speeds without gearing. Its disadvantages were all related to low power—short range, low maximum speed and long periods of frequent battery recharge. Nevertheless, for gracious city driving it was thought by many to be ideal, so much so that a few electric cars were being built into the 1930s. Electric cars have recently reappeared in small numbers, not for gracious city driving but rather in the main stream of automobile use.

A gasoline-powered engine creates power by introducing a gasoline-air mixture into the cylinders and then igniting that mixture with an electrical spark. The resulting small explosions drive the pistons downward or sideways depending on the engine's orientation, imparting mechanical energy to the crankshaft and thus to the transmission and then to the wheels. A diesel engine introduces an oil-air mixture into the cylinders. The high compression ratio of the diesel causes the mixture to compress so much that enough heat is generated to ignite the oil-air mixture, producing the same small explosion and application of power to the wheels.

The absence of an electric spark to ignite the diesel might seem to the uninitiated to be a large advantage because of the elimination of a complicated, and until recently often unreliable, electrical system of

---

12. See below, p. 25.

coils, capacitors, distributors, wiring harnesses and spark plugs. While it was, there was a price to pay for this simplification. When a diesel engine was running, there was sufficient heat remaining in the cylinders to ensure that the next burst of the fuel-air mixture would be ignited at the appropriate instant. On starting, however, the cylinders were cold and sapped so much of the compression-generated heat that the mixture would often not ignite. The solution, particularly in cooler weather, was to preheat the cylinders with electric heating elements (not nearly as complicated as the ignition circuitry of the gasoline engine). The solution took time to work, sometimes in particularly cold weather several minutes and never much less than 15 seconds. Not much you might say. Correct if you were starting heavy equipment that would be running all day once started. Very inconvenient in passenger cars in the eyes of many drivers, this being one reason why diesel cars had such low popularity, although low power was certainly a factor as well. The problem of starting diesels was so great and the fuel consumption at idle so low that diesel-driven railroad engines and construction equipment were occasionally left running overnight until environmental concerns inhibited the practice.

Now compare the effect of a maximum delay in starting diesel cars of not more than several minutes to a minimum delay of getting steam up of 5 or 10 minutes in a steam-powered car and you can see one of the reasons why, with all of its advantages, the steam car eventually failed. Efforts to revive the steam car in the late 1960s to take advantage of its much cleaner exhaust[13] and great power, used flash boilers for rapid steam, organic fluids in place of water in the boiler (among other advantages, they did not freeze or cause rust) and recirculation so the boiler fluids were not lost. There were very significant cost advan-

---

13. The combustion occurring in a steam car is external combustion, a seldom-used term but a common event. With combustion occurring in the open as opposed to occurring within a closed area like a cylinder the products of combustion are mainly carbon dioxide and water vapor. While carbon dioxide does contribute to the greenhouse effect, it is no more directly detrimental to people than a bottle of a carbonated soft drink.

tages over traditional steam engines but by then (i) it was probably too late to introduce so much new technology into the industry supporting automobiles, and (ii) the fuel economy was somewhat worse than internal combustion engines. Since the first OPEC doubling of the price of oil occurred in late 1973, much cleaner exhausts and simpler mechanics fell before fuel economy, dangerous exhausts and well-established technology and infrastructure.

In the 1890s many on both sides of the Atlantic were experimenting with automobiles. While the Benz car in Europe and the Duryeas' in the United States are often pointed out as the invention or the first practicable development of the automobile, in 1879 an American named George B. Selden filed for a patent on an automobile he invented in 1877.[14] The machine looked much like some other early cars except that the engine and the front axle were combined and could be rotated 180 degrees in order to reverse direction. This was a hard way to avoid installing a reverse gear but was somewhat analogous to the reversing of a horse and cart. Mr. Selden's patent was eventually issued and most of the automobile experimenters of the 1890s and manufacturers of the early 1900s paid royalties for the right to use the covered technology. At that point it would have been possible to say that the automobile had been invented, not developed, and that the inventor was Mr. Selden.

It was not that simple. Henry Ford, of whom more will be said shortly, refused to pay Mr. Selden's royalty, litigation ensued and the Selden patent was eventually found to be invalid, at least with respect to automobiles of the type manufactured in the twentieth century. Thus, while there were many inventors of parts for automobiles, there was no inventor of the automobile as a whole.

A number of problems had to be solved in order for the automobile to become a practical method of transportation. Two of the more difficult ones were determining the most likely type of power plant as outlined above and the method to be used to connect the power plant to

---

14. Clymer, *op. cit. supra,* p. 27.

the wheels which powered the automobile. These two problems were somewhat related.

Gasoline (and diesel) would seem in some respects to be the least likely sources of automotive power. The fuel was particularly dangerous and hard to refine, the engine was complex and difficult to start and the gasoline exhaust was particularly poisonous. In many ways, however, coupling the engine to the drive train could have been the biggest problem for gasoline engines. Unlike electric and steam engines, internal combustion engines cannot be started and stopped when the vehicle starts and stops, so a way to disengage the running engine from the stationary wheels is essential. Furthermore the power of internal combustion engines is very low at low engine speeds. But low speeds of the car are the times when maximum engine power may be needed. The maximum horsepower of such an engine is typically found at the high end of its speed range and the maximum torque in the midrange. Thus a complex system of gears was necessary to mate the engine's power to the automobile's needs and a clutch system of great strength was necessary to disengage the engine from the driving wheels when the car was stopped as well as when it was necessary to change gears. Such systems were developed, subjecting generations of drivers to the mysteries of manual gear shifts and clutches. It took half of the twentieth century before automatic transmissions became reliable and affordable, effectively eliminating the problem for those so wishing.

By the year 1900, the experimenters and tinkerers on both sides of the Atlantic began to give way to more serious attempts at practical automobiles. Perhaps the two extremes of practicability in the First Decade were defined by Rolls-Royce on the one hand and Ford on the other.

Rolls-Royce endeavored to build an effective, attractive, reliable car pretty much regardless of cost. Its early efforts of several years culminated in the 1907 Silver Ghost, a handsome, powerful, reliable, outrageously expensive machine that was well ahead of its contemporaries in concept and execution. It demonstrated to the world that automobiles could be so well built as to be durable, very quiet, very powerful, very well thought out from an engineering perspective and priced out of the reach of almost everyone. While Rolls-Royce demonstrated to the world what was possible if money were no object, there could never have been enough of their automobiles to create, by themselves, a demand for more and better roads as well as for more available service facilities.

By extreme contrast, Henry Ford had been building cars for most of the First Decade but his efforts were directed not as much at perfection but at production. To that end he experimented with different types of cars and different production methods. These eventually led to the first modern automobile production line in 1906 and, more importantly, the development of the Model T Ford in 1908. In sharp contrast to the Rolls-Royce of the same period, the Model T was ugly, noisy, of very simple and cheap construction, designed for owner repair and maintenance and was quite inexpensive. Both used gasoline-powered internal-combustion engines, although steam and electric power was still in the running throughout the First Decade and beyond.

Henry Ford was an incredible moving force (sorry) in the automobile world. He was not, however, an automotive genius from the beginning. Several of his early ventures in the field failed, but he learned well from his failures. Finally he introduced the Model T which differed from his earlier efforts in significant respects. First, it was a very simple car, most of the operations of which were not only understandable by a majority of the drivers but also were repairable by them. Secondly, the car was produced in a particularly efficient way with well thought out work flow and employee convenience. His production line even had subsidiary lines which, when the parts on those lines had been completed, merged with the main production line. The combination of these factors enabled Ford to produce his cars very cheaply, and he sold them cheaply. While he probably could have charged considerably more for the machines, he continually tried to find ways to reduce costs and thus prices. He felt that the volume increase would more than offset the reduced profit margin per car. Time certainly proved him correct. He also thought that the more cars that were sold the greater the demand would be for automotive-related services such as roads, repair shops, fuel delivery points and parking. With better services he felt that more people would want automobiles, a substantial portion of which would be his. He wanted his employees to be automobile owners both because in the abstract it seemed like a good idea and because Ford Motor Company had become so large as to be a substantial market for its own automobiles, having increased its market share by a factor of five from the year of introduction of the Model T to the beginning of World War I, and the market itself had expanded greatly in those six years.

Ford's well-known minimum pay of $5.00 per day (plus profit sharing), initiated shortly before World War I, had the dual motive of reducing employee turnover, a costly event, and making it easier for his employees to buy Ford cars.

The Model T, while revolutionizing the automobile industry, was not without some quirky aspects. Its height, combined with its primitive suspension, made it a very unstable machine. Fortunately, the bad roads of the period made it difficult to achieve even the low top speed of the Model T. The floor pedals of the Model T were, by present standards, peculiar. There were three. One was a conventional brake pedal although the brakes were mechanical, not hydraulic, and thus were not very effective or reliable. Another pedal was the forward gear shift; depressing it placed the car in low gear, releasing it shifted the car into high gear (there were only two speeds forward, sometimes described as slow and slower). The third pedal controlled reverse gear; depressing it engaged reverse, releasing it disengaged reverse. The accelerator or throttle was on the steering column. One slight advantage of this system was that panic stops were best accomplished by depressing all three pedals simultaneously. The results could be dramatic but hard on the machine.

Unfortunately, Henry Ford was too enamored of his creation. While Model Ts were being built all over the world and in large numbers, other manufacturers were making inroads into the market he had created. The Model T became cheaper (to under $230 from about $800 initially). It not only looked archaic but in many respects was becoming outmoded as the automobile industry evolved. Finally, in the mid-twenties, Henry Ford finally saw the handwriting on the wall and agreed to produce a new car. The Model A was the answer. It entered the market for the 1928 market year and was priced at about twice that of the Model T. The Model A was a more conventional car, at least by modern standards. It was powered by a much more powerful four cylinder gasoline engine and had a floor accelerator, a standard three speed forward and one reverse transmission with a floor-mounted shift lever, a usual clutch pedal and a gravity-fed fuel system with the tank immediately forward of the dashboard. The fuel gauge was a float in the tank seen through a glass window in the dashboard. The fuel flowed out of the tank, through the passenger side of the front seat on the rear of the firewall, through a shut-off valve[15] and then forward through the firewall to the engine compartment to a filter and the carburetor. And people thought the Pinto's fuel system was dangerous! Further increasing the danger of fire in the passenger compartment, consider the heating arrangements. Heat could be channeled into the right front passenger seat by a device that directed warm ram air over the exhaust manifold to the firewall in which there was a covered hole. With the cover open and the Model A moving at a moderate rate of speed, some semblance of heat could be detected at the cost of destroying the integrity of the firewall. To make matters worse (as if they

---

15. While having the shutoff valve in the passenger compartment created another potential source of gasoline leakage, it was necessary. Some drivers turned it off every time they shut down the engine. Because the fuel was gravity-fed to the engine, a leak anywhere in the fuel system would cause the fuel to leak out even if the engine were not running. The spark plug wiring was via bare copper strips which were prone to arcing to the engine block. The consequences of a pool of gasoline under the car vaporizing around the engine when it was started are obvious.

needed to be), an optional heater could be installed. That heater, located inside the passenger compartment, tapped a small amount of gasoline from the fuel line and ignited it. Much welcome heat was available at the risk of yet another source of a gasoline fire in the passenger compartment. Somehow most drivers seemed to survive. Nevertheless, like most early Ford designs it was simple and effective, and many routine services and repairs could still be done by the average driver including the author whose first automobile was a Model A which was two years older than its owner.

The purpose of the preceding paragraph was not to ridicule the Model A's fuel system or the car as a whole, for both the Model A and the Model T before it were very good cars for their eras. If one designs a car for significant owner repairs, the car will have to be as simple as possible and require few sophisticated tools. The result will often have to be a diminution of safety features. On balance it would appear that the designers of the Model A and the Model T made reasonable choices in this regard. The public tends to forget that one of the objectives in designing the Ford Pinto was to try to return to those simpler days. By that time, however, the mood of the automobile-buying public had shifted. Convenience and safety were more important than easy repairs and Ford was left with a somewhat dangerous machine that few wanted. It is not clear that the dangerous placement of the Pinto's fuel tank was part of the simplification process but it may well have been.

The Model A was a success. It was manufactured for five years and then replaced by the Ford V8 line as its premier product, but Ford's time of dominance of the United States auto market was over. It was now the era of the Big Three, Ford, General Motors and Chrysler. The time of any manufacturer's ability to rely on few models and on one color (you could buy a Model T in any color you wanted so long as it was black; Ford relaxed this rule with the Model A[16]) was gone forever. But none of the subsequent history could erase the enormous influence that Henry Ford's 1908 vehicle had on the world. Of course the millions of Model

---

16.   *See* Schild, *Original Ford Model A* (2003), *passim.*

Ts on the road in 1927 (Ford had built well over 15,000,000) did not disappear in 1928; they were still serving well in a wide variety of forms until World War II and even longer in other parts of the world where their ease of maintenance and ability in rough country continued to be essential. Only the Jeep of World War II fame was a worthy replacement, and it had an additional thirty years of automobile development plus the pressure of impending war to assist its designers.

The Jeep is an interesting example of something akin to the natural selection of evolution. Just before World War II the United States Army determined that it needed a small, light vehicle which could operate in rough terrain and carry 500 pounds of cargo. The Army sent its specifications to over 100 companies that it thought might be interested in competing for the contract. Only three responded and the winner was the Bantam Car Company which was as small as its name suggests. The two losing bidders, Ford and Willys Overland, were not only larger but very much larger. But Bantam, out of necessity generated both by time constraints and a lack of engineering personnel, had to scavenge for parts. It took an engine from another manufacturer, accessories wherever they could be found and adapted some parts for additional and unique uses. In the last category, for example, were the Jeep's headlights. Conventional headlights were, and still are, aimed ahead and slightly down in order to illuminate the road. By a slight change the Jeep's headlights were allowed to be moved manually from the usual position, not to follow a curve in the road as the Tucker and a few General Motor cars did later but rather to be turned backwards to shine into the engine compartment. By so doing, the original purpose of the headlights was preserved but an additional purpose was added, that of lighting the engine compartment so that repairs could be made at night in the field without any other source of illumination. Compare that to the natural evolution of the archaeopteryx's "wing" which was initially perhaps an aid to regulating body temperature and at most an aid to gliding but evolved into a real wing later[17] or to the

---

17.   Gould, *Bully for Brontosaurus*, pp. 139-51 (1992).

bat which adapted its social sounds for an additional use as a very effective sonar-like arrangement for navigating and feeding in the dark.

Parameters of the automobile having been set by Rolls and Ford, the gap between them was quickly filled with dozens of makers of cars and trucks. It soon became possible to buy cars with a wide variety of features at a wide range of prices. But putting those cars and trucks on the road created a demand for better roads. Before cars regularly ventured out of town, intercity roads were largely dirt, best suited for horses. The usual means of distance travel (over ten miles or so) was by railroad. This was not as inconvenient as it might seem with today's United States rail network. A hundred years ago the rail network in the United States was far more extensive and, perhaps more importantly, there was passenger service on almost all lines. It was possible to enjoy travel at a reasonable speed and in considerable comfort if one had the money to spend and the time to conform to the railroads' schedules.

As the automobile became more common and after pressure had been applied to various governmental authorities to improve the intercity roads, it became possible to travel on one's own schedule and not that of the railroads. At first it simply became easier for farm families to come into town, then it became possible for townspeople to reach other towns or cities more easily and finally it became possible to use automobiles for pleasure as much as for necessity. At each stage, as automobiles became more comfortable, faster and more reliable, and as roads improved the automobile became a more important part of twentieth century America, the mobility of the population increased in stages.

By 1940 this revolution in mobility had gone some distance, but long-haul roads designed for speed and limited access were just beginning to appear. The first section of Merritt Parkway in Connecticut (1938-passenger cars only) and the first portion of the Pennsylvania Turnpike (1940-all motor vehicles) had opened but the remaining economic effects of the Great Depression limited their use initially. It was somewhat surprising how few persons entering military service in the early 1940s had been more than 30 miles from their homes before induction. In that period few people moved very far from their area of birth. World War II changed that. Millions of those in military service saw greatly different parts of the country (and of the world, for that matter) as did those who moved, typically from rural areas to large cities, to work in defense industries. When the service personnel particularly returned to civilian life they were much more prone to mobility and much less tied to their roots, which mobility was further enhanced by the large number of veterans who took advantage of the educational benefits of the G. I. Bill and studied away from their former homes. The better postwar economy helped make cars more available and the country was then really on wheels. One of the more obvious indications of this trend was the rapid increase of "drive ins," particularly movies and fast food sites. Not all drive-ins had glowing long-term futures. Fast food is with us today, for better or worse, but drive-in movies seem almost to have disappeared. It is of some comfort to know that drive-in weddings ("you need never leave your car") apparently did not spread far from Las Vegas, perhaps because the laws of the United States would not permit drive-in divorces. The admonition of "act in haste, repent at leisure" was applicable here as in many other contexts.

The pressure was now on state and federal governmental planners to improve the long-haul road system. Many parkways and toll roads were the result, with significant improvement in the speed and comfort of interstate traffic but without a coherent pattern. The long-haul tractor-trailer rigs particularly benefited from this aspect of highway building. It was in this period that a rule of thumb developed to the effect that for non-perishable cargoes truck shipping usually was more efficient for distances of under 200 miles and rail was usually more efficient for longer trips. But big changes were coming.

When Dwight Eisenhower had been a young officer in the United States Army he was ordered to take an Army truck convoy across the country to test the feasibility of moving troops and supplies by road rather than by rail. It was a disaster. He never forgot the problems. When he became President in 1953, one of his objectives was the creation of a nationwide system of high speed roads. In 1956 the inter-state highway system was started. Eventually the country had a grid of fast, limited access roads with moderate grades and gentle curves, making it possible for passenger cars and trucks to move at the speed limit (or faster) most of the time. Automotive traffic thus was able to move at speeds limited largely by safety considerations, not road surfaces, grades, congestion or the like. The size of the country was now the primary determinant of the time long distance travel took and cars and trucks had pretty well reached their practical speed limits. But in the course of one century the United States had been transformed from a country with isolated pockets of population that typically moved very little from the place of their birth to an extremely mobile and fluid society, at least in a geographic sense. Social and economic mobility, while somewhat limited, had always been a feature of the United States to a greater degree than most countries. As a concrete example of what most can sense, there was a great increase in passenger cars in the United States, in 1900 approximately 8,000 passenger automobiles were registered and by 1999 the number was over 130,000,000. What may come as a surprise is that the number may be stabilizing. Slightly fewer private passenger cars were registered in 1999 than in 1990 and since 1985 the number has oscillated back and forth through quite a narrow range, although undoubtedly distorted to some degree by small trucks and sport utility vehicles substituting for passenger cars.

It is difficult to separate cause from effect but the improvement in the highway grid and the increased traffic load seem to have happened largely in concert. In the early fifties, the author made several long automobile trips from Pennsylvania across the northern states to the Rocky Mountains and across the southern states to south Texas and to Southern California. Almost all of the travel west of the Mississippi River, except within the western portion of Southern California where the California freeway system in its initial form was in operation, was on two-lane highways. In spite of that, traffic moved quite fast, including the large (for those days) semi-trailers. The southern route was the famous, or infamous, United States

Route 66, which, although it was generally subsumed by the interstate grid, does still exist independently in some places and is well maintained. The roadside businesses, restaurants, service stations and motels (or tourist cabins as they were sometimes described then) still exist as well. In driving some of the old stretches, still two lane, one can only wonder how we dared to drive it so fast fifty years ago, particularly since some of the western states did not have daytime speed limits so the only constraints were the maximum speed of the vehicle and the nerve of the driver.[18] The answer may be that we simply did not know any better and had not realized that there could be roads that were wider, flatter and with gentler curves that were far safer than the old two-lane roads or the even more dangerous three-lane roads, the latter now thankfully almost extinct. In some ways at that time the automobiles may have been more limiting factors than either the highway system or the traffic. There were still a few cars on the road with mechanical brakes and even the more common hydraulic brakes were not power-assisted as the are today. The engines were not particularly powerful,[19] were carburetor fed and were thus not comfortable

---

18.   The nerves of drivers could be tested on other roads as well. The original section of the Pennsylvania Turnpike (about 160 miles) was initially without an overall speed limit. A few drivers of high end automobiles (Dusenbergs preferred) tried to have the time stamps on their toll tickets less than two hours apart. While driving at an average speed of 80 miles per hour might seem dangerous but possible, that average speed had to include the transit of seven tunnels with 35 mile per hour limits and no room to pass other cars.

19.   But manufacturers were prone to exaggerate what power there was. The author was amused to learn, in the late forties, that at the beginning of World War II a giant manufacturer of electrical generating equipment was asked by the United States military to build a large type of portable gasoline-powered electrical generator for field use. The company calculated that a 100 horsepower gasoline engine would supply adequate power to meet the government's specifications. Since many automobile engines in those days advertised engine power in the 100 to 150 horsepower range, it seemed that it should have been simple just to buy a number of stock engines from an automobile manufacturer for use as the power source. Notwithstanding the sales force claims, none of the automobile engines actually tested to 100 horsepower; extensive modifications were needed to produce the necessary power for the generator.

with the great altitude changes experienced crossing the continental divide or the Sierra Nevada, but many of the drivers then were capable of making the necessary fuel mixture adjustments en route. Still, a coast-to-coast trip could be made in a week's or so of hard driving. Today, while the driving days would not be quite so long, the time would be comparable. Put another way, road design and construction and traffic seem to have moved, for whatever reason, pretty much together. The importance of all of this is that, at least since the end of World War II and probably a little before it, a person with a decent car could reach most of the United States from anywhere else in the United States in about a week, and this could be accomplished with reasonably good food and lodging on the way and no significant impediments at state borders. It is unlikely that any similar sized country would then have had available the same degree of mobility. This could only have a unifying effect on this country as well as providing the residents with many options of employment, education and lifestyle.

Obviously the automobile improved enormously from the First Decade to the present. Performance, comfort, reliability and accessories never dreamt of in the First Decade became routine even in bottom-of-the-line cars. This did not indicate a diversity of the First Decade automobile type development but only a dramatic improvement in the basic item. Even the offshoots, trucks, buses, tractors, fire engines, etc. were (i) minor modifications of the original automobile conception, and (ii) appeared very early on; the Model T alone in its many variants filled most of the possible functions. Thus the radiation of the "genus" automobile was slight and largely early in its life, with few deviations, almost all of which were both very closely related to the archetype and remained quite constant during the ninety-plus years since the end of the First Decade. Perhaps the recent "hybrid" cars, with their combination of hydrocarbon and electric power, will become a new "species" in the future.

It would appear likely that new power sources will soon be necessary for individual land transportation. A world-wide shortage of liquid hydrocarbons will probably occur. This could be accelerated by political or other complications in major oil producing areas, environmental problems or other as yet unforeseen issues or it could be postponed by more efficient oil recovery techniques, better fuel mileage and/or use of other sources of hydrocarbons. These sources, which are only slightly exploited today, could include oil shale (present in large quantities in the Rocky Mountains and tar deposits (found in large amounts in Canada and perhaps elsewhere). Eventually, however, another fuel source will be necessary and will have to be developed. Obvious possibilities would seem to be (i) petroleum derived from coal—the Germans did something like this in World War II with limited success, (ii) electricity produced by atomic reactors, and (iii) hydrogen produced by the application of the heat from atomic reactors to sea water. This last might be the most likely. Electrolytic reduction of seawater produces large amounts of hydrogen and oxygen plus small amounts of other elements, including metals. If the hydrogen is burned as fuel (and this has been done for years in testing internal combustion engines), the oxygen it uses in the process will about equal the amount of oxygen released into the atmosphere in the course of the electrolytic process producing that amount of hydrogen. Of course changes would have to be made in automobiles to accommodate this new fuel, the biggest probably being the fuel storage provisions. In order to store a reasonable amount of hydrogen, it would have to be liquefied. That could be achieved by storing it at very low temperatures, probably impossible in automobiles although it might be feasible for purposes of bulk storage. It could also be achieved by having it held under very high pressure in cars as well as in service stations. The transfer of fuel from storage to automobile would be more difficult than that of the gasoline pumps now in use but would in no way be impossible. Some of the complexity could probably be reduced by the use of fuel cells instead of outside free hydrogen. The space program has developed fuel cell technology quite far.

Ignored for this purpose is the possibility of powering automobiles with an on-board atomic power source, perhaps using an organic fluid in a type of "steam" engine, with the heat source being a very small atomic reactor or a very hot isotope. But the problems of incorporating adequate shielding and of accidental release of radioactivity seem overwhelming today. Similarly, electric power from thermionic emissions powered by hot isotopes presently seems unlikely. We should, however, always remember that yesterday's impossibility has often become tomorrow's commonplace. The future any distance away is always unpredictable and is usually surprising.

It was appropriate that Henry Ford became very wealthy as a result of his pioneering work in the popularization of automobiles. But it was even more appropriate in light of his continuous efforts to reduce the prices of his cars. This was hardly altruistic on his part. He thought he could make more money this way and he was right. The public, including Ford employees, benefited. It is also worth remembering that Ford, not having invented the automobile, did all of this without basic patent protection and in the process destroyed the principal patent by litigation. While not generally appreciated, a significant portion of the Ford legacy to the country was the huge charitable trust he created from the profits of the Ford Motor Company, the Ford Foundation.

The next century will probably see a reversal of the trend toward geographical mobility, at least by automobile. In urban areas the automobile has become so successful and the population so dependent upon it that it has become a nuisance in many areas. Public transportation, in a throwback to the railroads of the nineteenth century, should become more important. However, in the nineteenth century the railroads were new, exciting and such a great improvement over horse-based travel that, in spite of the fact that they were dangerous, there was not much resistance to their widespread use. Now the problem will be to convince the public to give up to some degree the comfort and convenience of the personal transportation which has so spoiled generations of travelers. Who knows what may happen but there surely can be no doubt that the simple devices of the First Decade were the precursors of the great importance of the automobile to the twentieth century and beyond.

4

Airplanes

We have no idea when human beings first thought about flying. Since writing is only about 5,000 years old and representational painting and carving about the same, we have no way of knowing for certain what earlier humans might have thought. It is not too difficult to imagine a neolithic hunter looking up enviously at a soaring hawk. By the time detailed sculpture and painting became common in the Middle East, winged gods and mythical figures were being depicted (for example in Egypt, the goddess Isis, and in Assyria, the lamassu). It is easy to see in these concepts that the ability to fly was a desirable attribute, so much so that it was reserved for gods or myths.

The first generally recognized "human" attempt at flight was probably that of Daedalus and Icarus, a mythical Greek father and son team who tried to fly by attaching feathers to their bodies with wax. It was said that the attempt failed when Icarus, in the best tradition of children, disregarded his father's advice and flew too near the sun. The sun melted the wax and the feathers came off, with the predictable disastrous results. Dedalus' aeronautical engineering, although innovative, seems quite dubious but the myth is a clear indication of a desire to fly if not of an ability to do so.

By the sixteenth century more serious thought was being given to the problem. Leonardo da Vinci's sketches show gliders, parachutes and helicopters but all are quite impracticable. It is likely that Leonardo was not alone in these thoughts at the time but his seems to be the only written material to have survived from that era. In the meantime the Chinese experimented with man-carrying kites. Writers throughout the ages have theorized and fantasized flight, of course.

The first actual planned lifting of human beings off the planet occurred in 1783 when the Montgolfier brothers, working in Paris, perfected a hot air balloon. The balloons were made of silk and inflated by hot air produced by straw burned in the basket below the balloon. This was obviously a dangerous procedure but it worked. Ballooning continued during the nineteenth century with hydrogen being substituted for hot air in most cases. While hydrogen produces more reliable and better lift, because of its propensity to burn it is also very dangerous. Since free balloons are at the mercy of the winds for their horizontal movement, they are rather impractical although some were used effectively in the siege of Paris in 1870-71. Tethered balloons on the other hand were useful in some ways for a time, mainly as spotters in military activities. Their usefulness was limited to times of clear weather and locations far enough back from the enemy forces to avoid being hit by hostile fire. Balloons could substitute for cavalry reconnaissance at times, having a wider field vision, a continuous rather than snapshot view and not usually being subject to enemy opposition. World War I saw the continued use of tethered balloons for observation purposes, but the danger to them had become more acute. Not only could long range enemy fire sometimes reach them but also the enemy's airplanes could attempt to shoot them down, usually by setting the hydrogen on fire. The risk to the balloons had become so serious that anti-aircraft guns were set up under the balloon and the observer was supplied with a parachute so that if the balloon could not be cranked down fast enough to save it he could simply jump out. Lest it be felt that tethered balloons were, so to speak, sitting ducks, because of the protective fire available around them they were considered such difficult targets that World War I fighter pilots listed balloons as well as enemy aircraft in their rosters of "kills."

Because balloons could not be directed, powered flight was still an objective. By 1900 it was more a dream than a reality although powered lighter-than-air ships were close to becoming practical. Conceptually, providing directing power to a vessel which can lift off without power is easier to visualize and would almost certainly require less power. On the other hand, pushing a large bag of gas through the air would be a slow process and the wind effect on the machine would be very significant. There was a complicated issue about the form of power to be incorporated. The engine had to be light, reliable and perhaps diesel fueled (to avoid having electrical sparks from the ignition circuit generated near a supply of highly explosive gas) even with the increased weight per horsepower as compared to gasoline engines. It was also discovered that helium could be used in place of hydrogen; it had somewhat lower lifting power but would not burn. Unfortunately for the rest of the world, by far the largest supply of naturally occurring helium was in the United States.

The lighter-than-air powered ship (an airship) was developed, and one could probably say invented, in 1900 by Count von Zeppelin in Germany with his rigid airships which became known as dirigibles. They performed well in World War I and though most of the thirties, but flammable contents, slow speed and vulnerability to weather, among other factors, contributed to their demise both literally and conceptually. Yet for a time in the twenties and thirties the dirigible seemed to have many advantages. While it was slower than contemporary heavier-than-air craft, the speed differential was not great, perhaps two to one on average, and the airship's very long range made up for some of the speed difference by permitting long flights with fewer fuelling stops. There was, for most of this period, no comparison between airships and aircraft in matters of comfort. Airships were roomy, quiet, long-ranged, smooth flying, luxurious and faster than ocean liner; aircraft were cramped, cold, drafty, very noisy and short-ranged. The two modes of passenger travel only began to become comparable in terms of comfort in the mid-thirties when the great flying boats, in particular the Boeing 314, came into service with their spacious cabins, sleeping accommodations and long range albeit at slow speed as compared to other aircraft. They maintained a good edge in speed over the airships, however, and might have had a longer term of use had it not been for World War II. The war eliminated the civil passenger flying boat activity over the Pacific Ocean and restricted it severely over the Atlantic Ocean. By the end of the war, the forced acceleration in aircraft development had produced faster, long-ranged land-based aircraft which could span both oceans with a minimum of refueling stops. The large passenger flying boats passed into history, although the concept of the flying boats still appealed to naval aviation planners because of their great endurance and they continued in naval service for a decade or more after the war.

As nationalistic symbols, it was hard to surpass the great dirigibles. They were huge, made enough noise to attract attention and flew relatively low making them particularly obvious to persons on the ground. The flying boats, on the other hand, while large as powered aircraft went, were not a great deal larger than other aircraft and thus not as attention-getting as the airships. Furthermore, because the flying boats needed large expanses of water for takeoffs, landings and emergencies, they flew almost exclusively over oceans where few would see them.

Of the various factors usually cited as the causes of the failure of the dirigibles, vulnerability to fire is typically thought to be the most important. There are good reasons for feeling this way. Hydrogen, the usual lifting gas, is intensely flammable, ignites very easily and there is a great deal of it. The only other lifting medium, the gas helium with newly-discovered supplies, was a virtual American monopoly. The Germans tried to negotiate with the United States for a supply and might have been successful had Hitler not been so ruthless in the mid-thirties that he discouraged the United States from wanting to deal with Germany about helium or anything else. Further limiting the attractiveness of dirigibles, hydrogen fires were also very newsworthy, both while in progress and in their aftermath. The pictures did not encourage ticket sales.

It should be noted, however, that United States dirigibles were lifted by helium, not hydrogen, and all but one, the *Los Angeles*, crashed. Weather was the cause. Rigid airships, for reasons of weight, do not have a strong structure and what strength they have is spread out over a large area. Weather, and particularly wind shear and thunderstorms, present high risks to airships even without the fire hazard of hydrogen. In aircraft we have only recently begun to master the handling of wind shear, or better its avoidance as we have been avoiding thunderstorms for years even though aircraft are far stronger and more maneuverable than dirigibles. Thus the argument that is sometimes heard, especially from Germans, that the dirigible would have been a great success if the United States had not hoarded its helium does not really conform to the facts.

The dirigible's cousin, the blimp, fared better. The blimp has no rigid structure and consists of a series of balloons contained in a soft outer container. It kept on flying for years after the last dirigible had been grounded, doing very useful patrol work during the World War II. Later its role was reduced to being a source of aerial views at sporting events and as a very large mobile billboard. Before writing off lighter-than-air ships as a complete failure, keep in mind that many persons crossed the North Atlantic Ocean nonstop in a lighter-than-air ship (the British R-34) before Lindbergh, and the R-34 made the trip in both directions. This was particularly important because the prevailing wind over the Atlantic is from west to east; as a result in something as slow as a dirigible the east-to-west trip takes much longer. Also in fairness remember, as too few do, that two Britons, Alcock and Brown, made a transatlantic crossing nonstop downwind in an old bomber eight years before Lindbergh and several weeks before the R-34 flights.

Still the main chance was the heavier-than-air craft. Some basic design work had been done in the nineteenth century especially by Carley and Stringfellow in England with gliders that looked surprisingly modern and actually flew. By the end of the century a number of persons were hard at work on much more sophisticated gliders and powered ships. The Wright brothers, Langley, Chanute, Lilienthal and others were trying to develop and quantify the concepts of lift, drag and thrust as well as coming up with appropriate control systems and power plants.

Apart from the Wright brothers, or perhaps even including them, Samuel Pierpont Langley seemed the most likely to succeed. He was a brilliant man, the Secretary (the highest officer) of the Smithsonian Institution and very well connected in Washington, D.C. He started his search for the secrets of manned flight by building models of biplanes eventually powered by steam engines. He was quite successful in designing models that would fly considerable distances although they traveled largely in straight lines.

His successful models proved that Langley had learned how to generate lift and apply power to flight. The models were, of course, not controllable in flight although he did incorporate dihedral in his wings, an important concept not used by the Wrights, perhaps because they did not appreciate its value. Dihedral is the slight upward sweep of the wings which provides some forces tending to return a plane to level flight if it rolls slightly to either side. This correction happens automatically because the lift generated by the wings is applied directly perpendicularly to the wings. If dihedral exists on a wing, the total lift of that wing is applied not directly vertically but is tilted slightly toward the fuselage to the degree of the dihedral, thus reducing slightly the component of lift being applied vertically to counteract the force of gravity. If a plane rolls slightly, the lower wing increases the lift applied against gravity as the wing approaches level and the lift component reaches maximum vertical effect. In the meantime, the higher wing's lift component shifts more away from the vertical and thus becomes less effective to hold the wing up. All else being equal, these factors tend to level off the plane and correct the roll, making for a more stable flight and easier control of the aircraft.

As a result of his successes with models, Langley received considerable funding for a full-sized, manned aircraft (which he called an "aerodrome"). He obtained a large internal combustion engine,[20] built a houseboat on the Potomac River for an attempted catapult launch of his creation, and failed ignominiously and very publicly in 1903. His failure was probably due in part to his concern for his pilot's safety. He thought that crashing in water would be less dangerous than crashing on land, which is not necessarily so. Once having opted for over water flights he had either to attach floats for water takeoff, with their weight and drag penalties, or use a catapult, with its severe stresses on the airframe. He chose a short and powerful catapult with disastrous consequences. Given the lack of structural strength of all early flying machines, it is odd that Langley thought his would be able to survive the force of his catapult without damage. Langley was ridiculed for his failure, did no further work in the field and died in 1906.

The Wrights in the meantime continued their work in relative obscurity in Dayton, Ohio, and on the sparsely peopled outer banks of North Carolina with no outside funding, no great prestige as scientists and little academic help.

Langley's prestige was such that the Smithsonian Institution claimed for some years that he had invented the airplane and efforts have been made, somewhat spasmodically, to prove that, but for X or Y or Z, the aerodrome would have flown. It did not and if somehow one had left the ground or the catapult it would have been uncontrollable in flight. The Smithsonian view was the cause of the Wrights having their machine displayed in England and not in Washington for some years. The Wrights were further outraged by the Smithsonian's later efforts to prove that Langley's machine could have flown by asking their arch rival, Glenn Curtiss, to build and fly a copy of the Langley aerodrome. He did, but the Wrights claimed he had made "dozens" of changes to Langley's design with the benefit of hindsight.[21]

---

20. See p. 51, fn 24.
21. Not until October 24, 1942, did the Smithsonian finally admit that the Wrights "were the first to pilot a powered airplane." Py-Lieberman, *From the Attic*, Smithsonian, December 2003, p. 46.

As is well known, the Wright brothers made their first four powered flights on December 17, 1903, at Kitty Hawk, North Carolina. The lengths of the flights varied from about 100 to over 800 feet; the aircraft was ground-launched, man-carrying, controlled and flew under its own power for almost a minute on one flight. What was the real significance of those flights?

It was not that the Wright brothers had discovered how to create wings that generated lift; a number of early experimenters had done that. The Wrights had done a great deal of theoretical and experimental work in the area and in so doing discovered that the leading technical material in the field, the Lileinthal tables, had to be corrected. Also importantly, the Wrights had developed a system of controlling an aircraft in flight. They installed rudders initially to stabilize the aircraft and later to control the motion of the plane about its vertical axis (yaw) and elevators to control motion about its lateral (wing) axis (pitch). Most significantly they devised a method to control motion about its longitudinal (fuselage) axis (roll). Lack of control in the air had caused many crashes of early pioneers.

Longitudinal axis control is very important in the turning of an aircraft. One might suppose that use of a rudder would cause turns. It can, but such turns are skids, disrupting the airflow over the wings and endangering the stability of the aircraft.

The Wrights recognized that turns were better made by banking the aircraft which required some change of form to the wings. Their solution was to bend the wings—bending down the rear edge would cause the wing to rise and the plane to turn away from the bent down wing. The Wrights called this technique wing warping and it was the primary basis of their patent for the airplane. The concept was copied from the action of the wings of soaring birds. Unfortunately for the Wrights, the birds had far more control over their wings than a simple twisting of them. Thus they could compensate easily for other related problems of wing warping that the Wrights could not handle well at the early stages. Warping the wings reduces the efficiency of the wings as lifting bodies unless other compensatory changes are made. Birds could make changes in theirs, the Wrights could not. The result was some loss of lift in the warped wing because of the warping as well as a result of the tendency of the aircraft to slip sideways into the lower wings, further distorting the airflow of the wings. Drag also increased. The results were control problems and the possibility of a stall. Neither were desirable, but stalls were particularly dangerous given how low the early planes flew and thus the short distance they had in which to recover from a stall. At this stage of aviation development the prospective pilots had not received the routine stall recovery training that has become an early element of the training of fledgling pilots. The good news was that (i) the distance to the ground was short, and (ii) the flimsy airplane itself was a bit of a cushion in case of a crash.

In due course the Wrights discovered that connecting the motion of the warped wing to the movement of the rudder at least began to approach what would become known as a coordinated turn, neither slipping nor skidding and with the "g" forces continuing to press the pilot squarely into the seat or, in the case of the early Wright models, into the cradle in which the pilot lay. While in any turn the effective lift of a wing is reduced (because, although the actual lift may not diminish, the lift is no longer applied exactly opposite to the effects of gravity to the extent of the bank), the airflow is not much disturbed in a coordinated turn.

In the meantime, others were trying to find better solutions, not only because they might in fact be better but also because they might thereby avoid the Wrights' patent. Glenn Curtiss, who seems to have discovered the Wrights' wing-rudder connection improperly, tried to work around the patent by building the usual biplane, but without the wing warping, and then installing small, wing-like horizontal control surfaces between the outboard sections of the wings. This system worked well, better than warping, and was presumably the basis for the name aileron (the diminutive of the French term for wing) now applied to the principal wing control surfaces. Over time, these wing

control surfaces returned to being an integral part of the wing but separately hinged so movement of them did not much affect the rest of the
wing.

The Wrights sued Curtiss for patent infringement. In the best tradition of patent cases, the matter dragged on for years but eventually the
Wrights won. The Curtiss and Wrights interests subsequently combined to form the Curtiss-Wright Corporation. In the operations of
that corporation what may have been to the Wrights' most overlooked
contribution to flight became paramount.

Not everyone agreed that the Wrights were first in flight. Beyond
the Langley supporters' claims and Curtiss' attempt to block the primary basis for the Wright patent, a Brazilian is celebrated both on their
postage stamps and in place names as the true inventor. Santos-
Dumont, a Brazilian by birth but working in France, had been working on powered lighter-than-air ships since very late in the nineteenth
century. From 1906 he turned his attention to powered aircraft. His
creations flew. His claim to priority was based not on the fact that he
flew before the Wrights but rather that his airplane flew without the
benefit of much if any wind whereas the Wrights flew into strong
headwinds. Of course the Wrights went to Kitty Hawk because of its
strong and steady winds. It is true that taking off into the wind causes
the aircraft to reach flying speed sooner than taking off in still air and
even a cursory glance at Santos-Dumont's planes suggests that their
wing loadings were much lower than the Wrights. The lower wing
loading would indicate, all other things being equal, that they would
take off at lower speeds. The size of the wings also suggests much
greater drag and therefore lower top speeds. Notwithstanding the argu-
ments on behalf of Santos-Dumont, with the exception of France and
Brazil the world seems to have accepted the Wrights as the inventors of
the airplane.

In their work on powered flight, the Wrights needed a power source
that was both light enough to be lifted by their aircraft and powerful

enough to move the propellers they had designed[22] with the necessary force. When they could not find a satisfactory power plant commercially available, they designed and built a four-cylinder inline[23] water-cooled gasoline-powered engine. It obviously did the job and may have been the most important new component of their first flights, although probably the least appreciated.

It is important to keep in mind that a large part of the Wrights' genius was combining components and ideas, many of which others had considered, into a workable complete aircraft. Thus their engine, while adequate for their machine, was neither the most powerful nor the most efficient in terms of power-to-weight ratio of contemporary engines,[24] but it was efficient enough when integrated into the rest of their creation.

The problem of maximizing power and minimizing weight has been chronic in the aviation engine industry. The history of Curtiss-Wright

---

22.   There were no practical aircraft propellers before the Wrights and ship propellers were not useful models. The Wrights had to design and build their own propellers from scratch. They seem to have been the first to realize that propellers with a wing-type airfoil work better because the propeller then both pulls the plane forward by the propeller's "lift" and pushes the air backwards for Newtonian "thrust."

23.   Inline engines have the cylinders lined in rows. They have a small frontal area and thus low drag. Since most of the inline cylinders are not able to be air cooled because they are behind one another and not in the cooling airflow from the plane's motion, a liquid coolant is necessary. Such a system works very well for cooling purposes but the liquid and its container are vulnerable to battle damage and mechanical problems. Radial engines have the cylinders arranged around the crankshaft, slightly displaced if there is more than one bank. The result is more frontal area per cubic inch displacement but less vulnerability because the cooling is by the airflow through the engine.

24.   The Wrights' engine at its most efficient had a power-to-weight ratio of about 10 pounds per horsepower and this ratio was only available for the very short period when 16 horsepower was developed; Professor Langley's engine supposedly had a ratio of about 2½ pounds per horsepower and produced 52 horsepower. Jakab, *Visions of a Flying Machine*, pp. 192-93 (1990). Two and one-half pounds per horsepower for an internal combustion engine seems suspiciously low for this period.

is perhaps illustrative of the relative importance of the corporation's origins. Initially it made both airframes and engines. The airframes were sold under the name of Curtiss and were early successes in the 1920's. The Curtiss Hawks (United States Army Air Corps model numbers P-1 through P-6, the "P" designating pursuit) were accepted by the Air Corps. They all used Curtiss inline liquid-cooled engines. Thereafter Curtiss had some successes with prototype airframes but only two pursuit ship contracts (the P-36 Hawk and the P-40 War-hawk) resulted in significant production. Its last contract, for the P-40 series, was a great success, although somewhat ironically the P-40 used an Allison V-1710[25] inline engine made by General Motors. With the completion of the C-46 (transports mainly powered by Pratt & Whitney radial engines), the SB2C (Navy dive bombers with Wright radials) and some Navy seaplanes (also with Wright radials), the Curtiss name died out after World War II. There were a number of Curtiss designed post P-40 prototype fighters up through the XP-87 Black-hawk in 1945 but none were ordered.[26] On the other hand, the radial engine business continued along well under the Wright name even though the Curtiss inlines were dropped. The size of the larger engines steadily increased from the R-1820 (one bank) to the R-2600 (two

---

25. The first letter in an engine designation describes the type of engine—V for an inline with two banks of cylinders meeting in a V, X for four banks in opposing V's, W for four banks side by side, R for radial, O for opposed, J for pure jet, T for turbojet and F for turbofan. The number (for non-jets) is the cylinder displacement in cubic inches and is thus roughly related to horsepower. Jet engines use a somewhat different description sequence.

26. Curtiss, of course, built other types of aircraft throughout its airframe history. The most notable prior to the Hawk series was the JN-4D ("Jenny") a World War I trainer with a Curtiss engine. After the war the surplus aircraft market was flooded with Jennys. Also in the Curtiss arsenal were such oddities as (i) the F9C ("Sparrowhawk"), a parasite fighter intended to be carried by airships which would both launch and recover it while airborne; the idea was dropped when the airships crashed only to be resurrected by others as the XF-85 to be carried by B-36s, also discontinued, and (ii) the C-76, an all-wood transport to be produced in case the United States ran out of aluminum in World War II.

banks) and finally to the R-3350 (two banks). There were occasional lapses in the string of successful radials, the most noteworthy perhaps being the R-2160 which had seven cylinders in each of six banks, or 42 cylinders in all. While the engine probably ran very smoothly, one could easily imagine the difficulty of cooling the last few banks of cylinders to say nothing of the maintenance problems of such a complex engine or of the comments of the mechanics trying to work on it. The R-1820 powered, among others, the B-17 and the C-47, the R-2600 the B-25 and the TBF, the R-3350 the B-29 and some civilian airliners. Wright jet engines were another matter. They were generally not particularly successful and ended, as a production matter, with the J65. It was surprising that the J65 was not more successful. Although it was a copy of a good Rolls-Royce engine, it seemed to suffer from efforts to increase its power and perhaps from American manufacturing procedures. Rolls copies had been successfully made by other manufacturers.[27] During its very successful radial engine phase, Wright engines competed with Pratt & Whitney on almost even terms. With respect to jet engines, in the early days Allison, Westinghouse, Wright, Rolls-Royce, General Electric and Pratt & Whitney all competed, but the first three effectively dropped out of the business as engines began to reach 10,000 pounds of thrust in the early 1950s. Nonetheless, perhaps echoing the Wright brothers early great contribution to aviation of the first effective aircraft engine, Wright radial engines still power hundreds of aircraft today. It is fitting that the aspect of the Wrights' pioneering flights which seems to have finally assured their success was also the part of their company that has lasted longest.

It is also fitting that the Wright family profited from the invention, although not as much as one might suspect. The Curtiss-Wright Corporation was, and is, a very successful business although it has branched from aircraft and engines to a wide variety of products. The

---

27. *E.g.*, the J48-P-5, manufactured by Pratt & Whitney (hence the "P"), the 48th jet engine considered by the United States military and the fifth modification of that engine.

Wright brothers had relatively little to with the commercialization of aviation. Wilbur died in 1912, before there was any great commercial activity and Orville did not participate to any significant degree except to the extent of his sale of his interest in what became part of Curtiss-Wright. His absence from that company may have been largely the result of his antipathy toward Glenn Curtiss because he had "stolen" their ideas or because he had, at the Smithsonian's request, tried to prove that Langley's aerodrome could fly. It might also have because management and organization were not the forte of the Wrights. They were tinkerers and inventors. Their genius was in their ideas and their implementing of them, particularly in their careful experimentation with various aspects of flight. It is certainly understandable that the Wrights were bitter about recognition not given them. Not only did they have the long fight with Curtiss, but the position of the Smithsonian on the primacy of Langley had to hurt. That they were honored more in England than at home could best be explained by Washington politics; logic had no place in it.

As the foregoing makes quite clear, not only did the Wrights invent the heavier-than-air powered aircraft, they and their successors continued to aid greatly in its development both in airframes and in engines. But in the United States there was a gap in that development between the middle of the First Decade and World War I. Part of the reason for the gap was a lack of interest by the United States government in aviation. While the government did eventually buy a few Wright aircraft and engaged the Wrights to train the first Army pilots, other countries, particularly Germany, Britain and France, seemed to be moving far faster toward significant air power. In naval aviation as well the United States lost the early lead it had when, at the very end of the First Decade, it made the first flights on and off ships.

World War I pushed aviation ahead rapidly. There had been some token appearances of the military use of observation aircraft earlier in Libya, in the Balkan Wars that preceded World War I and in the United States operations on the Mexican border beginning in 1914.

These flights, while sporadic, did indicate a probable use of aircraft in future wars. Soon, however, all parties realized that there was a bit more that aircraft could do in war. Early on in World War I the crews in observation planes would merely wave to each other but shortly they began to carry weapons to shoot at their enemy. From this informal beginning the fighter aircraft evolved. In the case of two seat aircraft, which was the usual configuration of observation planes, there came to be a flexible machine gun mounted in the rear cockpit. With it the observer had a field of fire to the rear and above. Another machine gun or sometimes a pair of machine guns would be fixed and firing forward for the pilot to use by aiming the plane at the target and the single seat fighter was born. The forward-firing guns presented a problem. If the aircraft were a pusher type with engine behind the pilot and the propeller behind the engine in a pusher mode, the guns could be mounted in the front of the fuselage with a clear field of fire. Pusher aircraft were rare, both for some aerodynamic reasons and to make the pilots more comfortable about their chances of bailing out successfully. In the usual tractor configuration, on the other hand, the guns had to be arranged so that they did not shoot off their own propeller. This could be done in two ways. First, the guns could be mounted on the top of the upper wing of the biplane fighters in order to fire over the top of the propeller arc, and later, when a device to interrupt the firing of the guns when propeller blades were in the way had been invented (generally attributed to Anthony Fokker, a Dutch engineer working in Germany, although it may have been invented several years earlier in Austria), the guns could be mounted on the fuselage in front of the pilot. Fragile wing structures prevented the guns from being mounted in the wings outboard of the propeller arc as was the case in the thirties and later. Aeronautical science and engine design progressed rapidly, spurred on by the exigencies of the war, and improved both bombers and fighters. Despite the romance that surrounded the various airforces, air activities were not a big factor in the war. Showing the level to which American air power had fallen, after inventing the airplane almost no United States designed aircraft served in a combat role in the war; its contribution was largely limited to pilots flying, generally, French mounts.[28] The only American plane in volume production during the war was a trainer, the Curtiss "Jenny."

Also serious thought had begun to be given to naval aviation although the only potential use of ship-borne aircraft could have been at the Battle of Jutland in 1916 when HMS *Engadine*, a seaplane tender, was part of the Grand Fleet. It was a great loss to the British that her observation capabilities were not effectively used for had they been there was a chance that the battle could have been a dramatic British victory. By the end of World War I at least three nations were preparing aircraft carriers equipped to launch and recover aircraft from large decks. Great Britain, Japan and the United States started down this path. There is some argument about which nation built the first aircraft carrier (as opposed to putting a flight deck on another type of ship). The Royal Navy deployed an aircraft carrier very early (HMS *Hermes*) but some purists would describe her as another adaptation and would credit the Japanese Navy with the first aircraft carrier (IJN *Shoho*). Both navies also had several conversions with flight decks by the early twenties. Whatever the merits of the British and Japanese claims to pride of place in the forefront of carrier development, it is quite clear that the United States Navy was a poor third. A surplus collier (USS *Langley*) had been converted for aircraft use by the installation of a full length flight deck but even the next two United States carriers (USS *Lexington* and USS *Saratoga*) were conversions from battlecruisers. The first carriers of the United States Navy to be constructed from the keel up as such did not begin to arrive in the fleet until the 1930s. The American conversions served better than the others once World War II started. *Langley*, which had been relegated to an aircraft transport, was sunk in early 1942, but *Lexington* and *Saratoga* served in the fleet during the war. *Lexington* was sunk in the Battle of

---

28.   The exceptions were a few Curtiss flying boats with British crews. Boyne, *The Influence of Air Power upon History*, p. 71 (2003).

the Coral Sea but *Saratoga,* while damaged, served throughout that war.

After World War I civilian aviation came into its own. Airlines were formed and began to offer regular if not comfortable daytime service. Barnstormers (freelance stunt pilots) brought aviation into the countryside. Even military aviation, after a substantial number of years of stagnation (because after all the Allies had just won the "war to end all wars" so warplanes were not likely to be needed in the future) began to improve slowly. By the mid-thirties passenger traffic in North America and Europe had become well organized and somewhat popular, helped by a group of passenger planes led by the early Fokkers and the Ford Trimotor and proceeding through the Boeing 247, the Douglas DC-3 and the Junkers Ju-52. The ability to fly passengers at night and in bad weather helped to popularize airline flying; better navigational aids did not hurt either. By the late 1930s even transoceanic air travel was possible, due to large, long-range, very comfortable flying boats made by Sikorsky, Martin and Boeing.[29] Thanks in part to the pilots trained for the war and to the romance associated with flying in that war, but probably the most influential force was a huge supply of training aircraft (the Jennys again) left over and sold very cheaply, flying had become a factor, although small, in the world economy.

As the possibility of another great war loomed in the early and mid thirties, military aviation leaped ahead. Naval aviation made particularly dramatic advances. Most first-line aircraft carriers were now designed from the keel up as carriers rather than being adaptations of other types of ships (cruisers, battlecruisers, colliers, etc.). Their aircraft, instead of having only the ability to take off and land on carriers, were now intended to be able to compete in large measure with land-based aircraft (always excepting the seriously non-competitive British Swordfish which in spite of, or some would say because of, its performance limitations performed surprisingly well). As war approached, training programs for aircrew, both army and navy, accelerated dra-

---

29. See pp. 43-44

matically, aircraft production by belligerents and prospective belligerents skyrocketed and new aircraft were designed and produced in record time. Perhaps the best example of the last was the P-51, a plane designed and built in under four months by North American Aviation pursuant to a British request. North American had never built a fighter before. With a different engine and some other smaller modifications it became probably the best propeller-driven fighter of the war. Other countries were also producing good aircraft rapidly, both in the design and production stages. Not to be ignored in a review of aircraft progress in the war were the vast improvements in aircraft engines during the same period. Before the war most reciprocating military aircraft engines were in the 700-900 horsepower range. By the end of the war they were approaching 3,000 horsepower and the weight per horsepower had dropped as well. That is not to say that all of the problems with aircraft design were solved during the war. The speed of sound had become a serious problem. During the war, as fighters became faster, compressibility issues as they approached the speed of sound became serious. Supersonic flight was finally achieved in 1947 but even then the last major step in understanding transonic flight, the area rule, was not articulated until the early 1950s.

Between the wars there had been a good deal of thought given to the use likely to be made of air power in the next war. Most planners felt certain that whatever use was made, it would be much more than was the case in World War I. Planes would be much faster and fly higher (and thus be harder to shoot down), would have much greater bomb loads (and thus could do more damage per plane), would have longer range (and thus could have bases further back of the battle lines and reach further into enemy territory) and would be better armed. It was also felt in many circles that, whatever the defenses, the bombers, or most of them, would always get through to their targets. If that were to be correct, the appropriate reaction would be to establish a counterbalancing offensive bomber force much as was done during the Cold War with the mutual destruction strategy, although in the former case the weapons to be used were obviously very much less powerful. In any case, it was felt that aircraft would be a most important factor in the next war.

In the event, the results of strategic bombing in World War II were very different than anticipated. The bombers did almost always get through to their targets although often at great cost to themselves. The bombing results were usually disappointing, at least to the attackers. The devastation was not as predicted for a number of reasons. First, the weather conditions were seldom ideal for daylight bombing. Over Western Europe and England there was often cloud cover and over Japan there were usually high winds. Second, the bombsights were generally not very accurate (with the possible exception of the American Norden bombsight) and, being optical, were worthless for bombing through clouds and not particularly effective at night. Third, the bombers were, for most of the war, not able to defend themselves adequately from either anti-aircraft fire or enemy fighters. The British appreciated this last point from the beginning of the war and preferred to bomb at night because for most of the war German night fighters were not big factors. Of course bombing accuracy was worse at night. Even the United States, with bombers more ruggedly built and more heavily armed at the expense of a heavier bomb load, discovered that daylight raids into Europe required long range fighter escort which was not available until 1944. Without that type of escort, losses on long range missions were unacceptable.

As a result, the fact that the bombers usually got through did not validate the offensive strategy because the bombers seldom destroyed their targets. Additionally, in Japan the decentralized industry meant that there were often not even obvious targets.

One thing lead to another and by the later stages of the war the United States and Great Britain emulated the German night blitz on England in 1941-42 and all often bombed areas rather than precise sites. Even the German V-1 and V-2 weapons in 1944, while deadly, were woefully inaccurate. Hundreds of thousands of civilians were killed or wounded from the air in England, Germany and Japan. And thousands of airmen were killed, wounded or captured. The United States Eighth Air Force, the principal United States bombing force based in England, lost about 25,000 men killed and a like number captured; the Royal Air Force Bomber Command lost twice as many.[30]

Consider what this meant in terms of the human cost of war. Hundreds of years ago, most wars, at least in Europe and North America, were fought by relatively small armies with not terribly effective weapons. Many battles were fought as a matter of maneuver. A general outmaneuvered would often order a retreat; the army that ended the day in possession of the battlefield was deemed the victor. Napoleon changed the rules. He ordered conscription, marshaled more resources and fought battles of annihilation, understandably because he was confronted by enemies and potential enemies in various directions so he had to try to destroy any enemy he could reach. After the Napoleonic wars, weapons, particularly weapons used in a defensive mode, improved. Cannons now fired explosive shells, not solid shot; rifles were more accurate and longer ranged; and barricades began to appear and be used effectively on battlefields. Astute observers of the American Civil War and a few generals in it realized that the defense now had significant advantages. But most military strategists did not take notice and continued to use the tactics of Napoleon and Wellington in the face of changed facts. The casualties caused by such tactics were enormous. Weapons continued to improve, machine guns, barbed wire, much improved artillery, better and simpler rifles which could be reloaded rapidly while under cover and better earthworks. But the generals were still awed by Napoleonic tactics and the huge infantry losses of World War I resulted. Some optimistic souls hoped that more efficient air power in World War II would break the infantry stalemates with far less loss of life. Dr. Gatling had the same altruistic thoughts when he invented the Gatling gun in the mid-nineteenth century. His gun, an early and bulky but quite good version of the machine gun,

---

30.   Neillands, *The Bomber War*, p. 379 (2001).

was hoped to be so deadly that no army would use the close-order frontal attacks that had produced such high losses in the past. This was wishful thinking as even more effective machine guns mowed down infantry charges in huge numbers during World War I. Those who viewed air power in the same way were equally in error.

There was another element of the effect of air power on casualties in World War II. The majority of the military casualties in that war were caused by artillery. With the exception of the Russian-German front, military casualties in World War II were lighter than in World War I and perhaps the difference was caused in some measure by air power, However, civilian casualties, mainly in England, Germany and Japan, were much heavier in World War II than World War I and virtually all were caused by air power.

There is no doubt that the seers were correct. Air power was somewhere between very important and determinative in the ground portion of World War II. In the naval portion of the Pacific Ocean war, naval air was obviously determinative.

Although the post World War I era was one in which aviation was popularized, the post World War II era was when it was commercialized.[31] The developments of that war were rapidly incorporated into civilian aircraft. Such improvements as more efficient, comfortable and cheaper transports became available. Initially these were simply military cargo aircraft with new paint and better seats. Military standbys like the C-47 (DC-3), C-54 (DC-4) and C-46 were the first stage but soon the technology of the war years, particularly in engines and electronics, was used in a second wave of early post-war aircraft such as later models of the Constellation (the initial model of which was the C-69 used by the military in the latter part of World War II), the Convair and Martin smaller two-engine transports and the DC-6.

---

31. The author was a beneficiary of post World War II aviation, having served as an aerial navigator, pilot and weapons operator in military and civilian aircraft in the 1950s.

The private plane market was similarly aided by the war. Light planes could now have the electronics and instrumentation that even airliners could not have enjoyed before the war. Furthermore, there were now more companies able to build aircraft than before the war because of the necessity of farming out production of aircraft and components during the war to companies that had not previously been in the aircraft business.

Not to be overlooked in viewing postwar aviation were the vast number of trained aviators produced during the war. Not only pilots but other aircrew and ground crew had been introduced to, and predictably many wanted to stay with, aviation. In addition, the G. I. Bill allowed Americans who were otherwise qualified to upgrade their aviation credentials at little cost. Also contributing to the post-war aviation boom were the large number of airfields constructed by the military. Most of these fields now made surplus were training fields and were located in areas with good flying weather. The South and Southwest thus benefited most although many of the fields were located in desert-type areas because the large land areas needed for fields could be obtained there more easily and cheaply. These were areas that tended not to have many people living in them after the war and this caused the post-war aviation boom to be somewhat diminished.

In spite of the foregoing, the private aviation business did not explode after the war as many had predicted. While there was an increase initially, it was not as dramatic as some had thought. Aviation communities with light planes in the garages and driveways leading to runways did not spring up around the major cities, the convertible car/plane was never made practicable and the concept that many businesses would have their traveling personnel, salesmen, trouble shooters, etc. trained as pilots with their own light planes never really caught on. Costs and pilot proficiency were the usual limitations, although the speed differential between light aircraft and airliners, especially after jets arrived for the latter, reduced the perceived efficiencies of light planes flown by traveling employees and the improved highway grid made automobile travel more effective than it had been in the past. Of course, in some sparsely populated areas, like Alaska, Southwestern portion of the continental United States, Australia and Africa, the light plane revolution worked. Nevertheless, the post World War II civil aviation industry was a commercial success, particularly in the areas of long haul passengers and freight.

Post World War II aviation differed from post World War I aviation in another aspect. Military aviation development did not stop or even slow down after the war. Because of the Cold War and sporadic shooting wars, the Soviet Union, Great Britain, France and the United States continued to advance the design of airframes, engines, avionics and weapons systems. The last, of course, had little effect on civilian aviation but the first three, and especially engine evolution, produced major changes in airline travel as well as, later, in private aircraft. The jet engine, appearing operationally at the end of World War II, had the biggest impact. Contrary to popular opinion, jet engines were not a new idea which appeared toward the end of the war. Frank Whipple had patented a jet engine in England in 1930 but it was a long way from being a practical form of aircraft power. Absurd as it seems in retrospect, there was not enough interest in the project to finance development until the late 1930s. In fairness to those persons, military and civilian, who were not enthusiastic about jet engines in the beginning, jets were not fuel efficient, they were hard to airstart, prone to mechanical failures and foreign object damage and generally took a great deal of getting used to. The Germans and the British had created jet engines just before the war and were working on them throughout the war. Even the United States, which lagged seriously behind the Germans and the British in this regard, had issued a contract for its first jet fighter (the Bell P-59A) before Pearl Harbor and flew it the next year but with British designed engines. Unfortunately it was not a very good aircraft; the second American jet fighter (the Lockheed P-80) was, by contrast, very good and was available right at the end of the war. In the meantime both the British and the Germans used their few jet fighters operationally in the last stages of the war.

After the military had improved jets, and particularly the reliability and fuel consumption problems of jet engines, civilian jet transports became a possibility. These transports came in two categories, pure jet and turboprop. The former simply used jet engines directly; the latter used jet engines to drive propellers. The British won the race for commercial development with the Vickers Viscount turboprop and the Comet pure jet. The Viscount was a success in terms of passenger comfort and safety; it was very popular with passengers because of its smooth quiet ride. It failed economically because its operating costs were so high that it needed almost a full plane on each flight. The Comet had a serious design flaw which caused several crashes and, while the flaw was remedied, the aircraft's reputation never recovered although the Royal Air Force did use the improved Comet (as the Nimrod) for patrol purposes. The American jet situation was exactly the reverse. Its first turboprop, the Lockheed Electra, also had a serious design flaw which caused crashes. The flaw was corrected but the travelling public never accepted that the aircraft was then safe. The plane was adapted by the United States Navy as a patrol aircraft (the P-3) and is still in service. On the other hand, the first American jet transport (the Boeing 707) was a significant success, helped greatly by the extensive work that Boeing had done for the B-47 bomber and the KC-135 tanker, both very effective aircraft which contained a number of significant innovations.

The impact of commercial jet aircraft can hardly be overstated. In large countries, Canada, the United States, the Soviet Union, Brazil, Argentina, Chile, China and for transoceanic flights the advantages were great. Not only was travel time cut almost in half but because of the smoothness of the engines and the high altitude at which jets flew best, the ride was very comfortable. It was also surprisingly trouble free in spite of the apprehensions of those of us used to the quirks of early military jet engines which were run at their maximum speed in those applications. Those same engines were used in civilian aircraft but the derating (essentially a reduction in the maximum power) of those engines seemed to improve reliability dramatically. It would be quite fair to say that business and vacation travel were revolutionized by the ability to fly so far, so fast, so comfortably and so cheaply.

It would also be fair to say that but for the development of the airplane over the first half of the twentieth century the space program of the second half could not have happened. While it is far too soon to assess the results of the exploration of space, all indications are that those results will be very important. Even if they are not, the contributions of the technology developed for that exploration have already been felt. One important aspect of that technology relates more to radio and its progeny and is discussed in Chapter 7.

In contrast to the automobile development, there was a more or less continuous radiation of related, close and remote, flying machines. Initially, while many of the early machines looked quite different from one another, all powered flight fell into one of two "species," heavier-than-air and lighter-than-air. The latter, after a brief but spectacular flowering from the First Decade to the late thirties, declined to a mere but very large shadow of its former self limited to advertising blimps. On the other hand, the heavier-than-air group radiated rapidly from the thirties on. Considering conventional land and water aircraft as one "species" limited the radiation at first, but after 1930 new "species" began to appear at a slow but steady pace; autogyros, helicopters, jets, supersonics and vertical takeoff and landing aircraft. Even as the huge volume of aircraft made for World War II began to be converted back to the aluminum pots and pans from whence they came so that the total number of aircraft in the world was falling rapidly, the number of new "species" was increasing. The aircraft "genus" seems quite healthy, to the surprise of no one.

Under all these circumstances, it seems obvious that the Wright brothers flimsy, short-ranged, under-powered invention of 1903 was epoch making.

5

Atomic Energy

So far the First Decade events, whether invention or development, have been ones which most would view very favorably. Of course there will always be those who long for the "good old days" without the noise of railroads, the traffic of automobiles, etc., or who view what most would call progress with a jaundiced eye for fear that some cherished institution or condition would be adversely affected. Perhaps one extreme example of the latter was a First Decade objection to the introduction of electric trolleys and gasoline automobiles in place of horse drawn vehicles. The objection, surprisingly, was not to noise or pollution but was rather that the absence of horses would seriously harm the sparrows that thrived on grain in the horses' droppings in the streets. Fortunately no attention was paid to the objection, trolleys and automobiles prevailed and, to the surprise of practically no one, the sparrow population seems to be doing just fine on a somewhat different diet.

The next First Decade development to be discussed was, and continues to be, far more controversial. That development was the chain of discoveries that suggested the likelihood of nuclear chain reactions in turn indicating the possibility of atomic explosions and/or atomic power. It is difficult to pick a precise point at which this development might be thought to have begun. A case could be made for a series of events beginning with electron theory in 1895 (Lorentz), the discovery of uranium radiation the next year (Becquerel) and Rutherford and Bohr's work on atomic structure between 1911 and 1913. One might also place the beginning as late as 1939 when Hahn, Bohr, Einstein and others really got down to the basics of uranium fission.

But probably the best starting point is Albert Einstein's publication of his Special Theory of Relativity in 1905 even though it was somewhat collateral to, rather than in the direct line of, the discoveries. It is of importance in the context of nuclear fission in two respects. First it involved what the jargon manufacturers of today would call "thinking outside of the box," a concept that became more important in the field later. In fact the Special Theory was so far outside of the box it might have been called "off the wall." It was not until about fifteen years later that Einstein's Special Theory and his far more complex General Theory of Relativity were, to the surprise of many, proved empirically by astronomical observations, and atomic physics headed off in a new direction. Furthermore, illustrating the relationship of relativity to nuclear fission, in connection with his work on relativity Einstein derived the now famous $E=mc^2$ equation,[32] which was perhaps the most important aspect for it quantified the great power potentially available from the conversion of very small amounts of matter in the course of nucleus-splitting events. Research continued through the 1920s and 1930s, reaching a point in the late 1930s when many of the great physicists of the world felt that it might be possible to use spontaneous nuclear fission to create very rapid chain reactions in the U235 isotope of uranium, which reactions would culminate in an atomic explosion of tremendous force. It was at this point when Einstein wrote his famous letter to President Roosevelt, pointing out this possibility and urging the United States to proceed in that direction. The situation became even more acute when plutonium, another element with a fissionable isotope (Pu239), was discovered in 1941 (Seaborg and others). Germany too was working toward this goal. After an intensive and expensive effort (the Manhattan Project) which cost about $2,000,000,000 to develop the bombs and a related project cost-

---

32. "E" represents the energy released, "m" is the mass lost in the fission and "c" is the speed of light. While the mass lost is miniscule, the speed of light (300,000 kilometers per second) when squared is a huge number and the product of the two represents significant energy generated.

ing about $3,000,000,000 more for the design and production of air-craft (the B-29) to deliver them, the United States produced two types of workable atomic bombs which in fact ended the war with Japan. The next half century saw nuclear bomb capabilities developed by more than half a dozen countries, with the realistic possibility that one or more of the "nuclear club" would use a bomb to settle some dispute but, as it turned out, no one was sufficiently motivated or insufficiently aware of the possible consequences to make a first strike which was likely to generate a nuclear war affecting the whole world. While the risk certainly still exists in some parts of the world, tensions today seem much reduced overall from the height of the Cold War.

The reduced tensions certainly do not eliminate or even reduce the possibility that some nation, or group, would use an atomic bomb for some political purpose. So long as there are effective weapons that become available in one way or another there is always the possibility of their use. But that type of use is most likely to be a surprise without a period of tension. In the prolonged period of tension between the Soviet Union and the United States from the late 1940s to the early 1990s and particularly during the ballistic missile era (the late 1950s to the early 1990s) the varying political tensions were compounded by the very short time available for decisions about responses. From the time that an intercontinental missile rose high enough over the Soviet Union to be detected until it would explode in the United States was 25 to 30 minutes. During that period the United States would have had to have detected the missile, determined its impact point and assessed the degree of threat. The last may strike one as absurd. With evidence of an incoming missile presumably carrying one or more nuclear weapons toward the United States, why would anyone hesitate to launch the missile response and scramble the manned aircraft response? The latter can be recalled but the former once launched is unrecoverable. It might be possible to disarm the missiles after launch but at the very least the missiles will be lost. A single missile would have been a most unusual form of attack in that era. While it might indeed be an attack, it might also be some sort of radar glitch or a Russian error. This sort of dilemma not only delays response, it puts tremendous pressure on the decision maker. The decision maker might be inclined to make the "safe" call and launch the response. Living with this kind pressure on either side is extremely stressful. Lest this be thought to be merely after-the-fact rationalization, the author is aware of at least two instances (and there surely were more) in which responses could have been started but cool heads controlled the possible disaster. Forty years of this type of confrontation without serious incident was indeed remarkable. However, the present international climate makes it more likely that an intentional single launch could occur.

The basic platform for the creation of atomic bombs by itself is more than enough to justify the inclusion of the Special Theory of Relativity as a small event in the First Decade which spawned enormous consequences for the rest of the century and beyond, potentially affecting all life forms on the planet.

There is, of course, the other side of atomic reactions. If the reaction can slowed down and well encased to prevent radiation leaks, it can provide a great source of heat. If that heat is applied to water or some other liquid, the vapor pressure created can be used for many purposes. At present the two principal purposes are as fuel for naval vessels and for electric generating plants. The latter provides the promise of virtually unlimited electric power if the atomic reactors can be made safe. However, the history of electric power plants is not entirely comforting. Chernobyl and Three Mile Island are still in the public mind although a nuclear engineer acquaintance of the author feels that Three Mile Island illustrates well that the automatic safety systems can contain the consequences of human error. Even if correct this is not a complete endorsement of the safety of well-designed atomic power plants. Naval use, at least in the submarine service of the Soviet Union, is replete with nuclear accidents and fatalities.[33] Some of those may have been the result of operating errors but probably most can be attributed to the hasty design, limited safety backup and rapid construction undertaken by the Soviet navy in its efforts to catch up with the United States Navy which had taken an early lead in nuclear propulsion for ships.

In terms of the natural history analogy, it might be fair to view the atomic reactor development as falling into three "species." First, of course would be the atomic bomb, three of which were in fact exploded in 1945, one in New Mexico as a test and two over Japan. After World War II many additional atomic bombs were built and some tested. Two additional species soon appeared. One was the hydrogen bomb, differentiated by its much greater power and its very different construction. The hydrogen bomb uses an atomic bomb as an igniter and receives most of its power from the fusion of hydrogen not from the fission of uranium or plutonium. As such it does seem to deserve a separate "species" category. There are at present thousands of atomic and hydrogen bombs stored around the world, largely by the United States and Russia with another probable half dozen nations having at least one.

The far more benign "species," the controllable atomic reactors, have appeared in many countries, mainly as electric power generators but in the United States and Russia additionally as power plants for ships. The United States, after one unpromising effort with a merchant ship (the NS *Savannah*, named for the ship that in 1819 made the first transatlantic voyage at least partially under steam), limited its seagoing reactors mainly to submarines and aircraft carriers. The application of atomic power to submarines in place of diesel and battery power has enormous advantages. Not only does the submarine avoid surfacing (or near surfacing if a system for sucking air down into a submarine just

---

33.   Huchthausen, *K-19; The Widowmaker-the Secret Story of the Soviet Nuclear Sub-marine* (2002), and particularly the appendix of Soviet submarine nuclear accidents and incidents.

under the surface is used) to recharge its batteries from internal combustion engines needing air for operations, it has as much power
underwater as on the surface. It also has range and underwater endurance limited only by the on board food supply since methods have
been found to recycle and purify the air supply. The need for atomic
power in aircraft carriers was not so obvious nor so pressing. Even with
atomic propulsion, carriers need refueling for their embarked aircraft as
well as provisions for a crew of 5,000 or so. Thus unlimited fuel for the
ship's engines, while obviously helpful, does not make the carrier as
independent as the atomic submarine. None the less, the first atomic
powered aircraft carrier, USS *Enterprise*, was such a success that after a
period of proving its reliability in the fleet all subsequent United States
aircraft carriers have been nuclear powered. The Soviet nuclear submarines have been a serious problem for the Russian navy which inherited
them. They are very expensive to operate, require well trained crews
and are difficult and expensive to dispose of. Both nations have used
atomic power in other types of vessels on occasion but that was rare
and the other applications seem not to have been very successful.

6

HMS Dreadnought

The previous three chapters have dealt with the development of the bases for atomic energy, the invention of the airplane and the making practical the automobile as major technological events of the First Decade with enormous impacts on the rest of the century. This chapter deals with a British battleship, HMS *Dreadnought*. It may seem a strange choice since, while many battleships were built for and used in the World Wars, only a few battleships were in service after World War II and those only served sporadically for the next twenty-five years. By the end of the century the few battleships which still existed were only static memorials to events or to the men who served in them. A better battleship hardly seems in the same league with the creation of the automobile, the airplane or atomic energy. Bear with me, for this particular battleship may have had an extraordinary effect on the whole century and beyond, a remarkable feat for a naval vessel that seldom fired her guns in anger. Her total combat activities were as follows: On August 8, 1914, she fired at what was thought to be a submerged submarine off Scapa Flow and on a number of occasions in 1917-18 she fired at aircraft. Her most important action did not involve her reason for existence, her heavy guns. On March 18, 1915, she rammed and sank the German submarine U29, the only battleship ever to sink a submarine.[34]

A history of the battleship is a necessary preamble to this chapter. First it must be realized that the term battleship as it is known today is

---

34.   Roberts, *The Battleship Dreadnought*, pp.17-18, 20-21 (rev. ed. 2001).

a bit of a misnomer. Any naval vessel which is designed to fight other vessels is logically a battle ship. But the term battleship evolved differently.

In the days of organized fighting sail (say, 1500 to 1870) large ships, generally three-masted and rigged largely with what were called square (actually rectangular) sails, became the major shock force of navies. For several hundred years, from sometime after the defeat of the Spanish Armada by the English fleet in 1588, until the end of World War I and in a few instances in World War II, battles between the largest naval vessels usually took place with those vessels sailing in more or less parallel lines. Because until the latter part of the nineteenth century most of their cannons were located along the sides of the ships, fired outboard and had very limited and difficult possibilities of traverse, sailing in a line ahead (the battle line) and firing more or less at right angles to the direction the ship was sailing brought the most gunfire to bear on the enemy. It was called broadside fire for obvious reasons. Because every ship had a few guns pointed forward and aft, usually of smaller size than the broadside guns, a full broadside at best had only somewhat less than half of the ship's total weight of shot in it. Of course, if one found oneself between two enemy ships both broadsides could be used, firing almost the full weight of shot but presenting other problems. Being shot at from both sides was not usually felt to be desirable and few ships had large enough crews to man both broadsides and work the ship simultaneously. Only those ships that carried the largest cannons and were the most strongly built were deemed fit to serve in the battle line, and they were called line-of-battle ships. That designation was so delineating that it was often considered dishonorable for a line-of-battle ship to fire at smaller ships engaged in the same battle unless the smaller ships fired at it first. It was from that designation that the term battleship derived.

Once battleship design had coalesced into a three-masted, square-rigged vessel with two or three gun decks, little changed for several hundred years. The ships were wooden, the guns muzzle loading, smooth bore and firing cast iron solid shot usually weighing from nine to thirty-two pounds per round. While the guns had a range of two to three miles under ideal conditions, the lack of accuracy of the smooth bores (because the shot wobbled in the barrel and went out without any particular spin on it) and the inherent low trajectory of the guns dictated that most battles be fought at very short range (20 to 60 yards) in an attempt to smash the enemy ships, to dismount their guns, to take down their masts and to kill or injure their crews from direct contact with the shot and from large splinters of wood thrown about by the solid shot hitting the wooden structure of the ship.

During this period, stability of design was paramount. Battles were won or lost largely on the basis of the number and strength of ships one country could pit against another as well as the comparative tactics and ship and gun handling ability of the participants. Of course there were a few changes in design and weapons. Perhaps the most notable of the former were the replacement of a lateen yard on the mizzenmast by a gaff-rigged fore-and-aft sail (a spanker) in the mid-eighteenth century and, early in the period, the invention of a wheel to control the rudder. The rudder had previously been moved by an awkward system of levers from the quarterdeck or other steering location down to the rudder. In weapons, apart from a gradual increase in the bore as it became possible to produce larger gun castings and efforts to improve the reliability of the ignition, the only significant change was the invention of the carronade in the latter part of the eighteenth century. This was a lightweight cannon which fired a much larger shot than conventional ship cannons, often twice the weight of the conventional load. A lighter weight cannon firing a heavier shot might seem a contradiction but the explanation lies in the size of the gunpowder charge in the weapon; a carronade could only throw its shot a short distance although at that distance it was extremely effective. The appearance of the carronade created severe tactical problems. A ship armed largely with carronades was deadly at close range but was almost helpless at long range; a ship with largely conventional cannons was much more powerful at a dis-

tance. Naval strategy demanded that ships be placed in battle forma-
tions consistent with their structure and their armament. The object
was easy to state but sometimes difficult to execute. While this concept
was a tactical consideration in sea battles it became even more difficult
when a significant number of carronades were involved because of the
characteristics of the cannon.

The nineteenth century was a very different matter in terms of battleship innovations. First, a colonel in the French army invented a
form of exploding shell which could be fired from a conventional cannon in lieu of the usual solid shot. Previously the only similar weapon
available was a mortar which lobbed exploding shells for short distances at the expense of a very inefficient sail rig on the firing vessel to
keep any forward sails away from the departing shells. Mortar shells
were very effective when accurately fused and fired but setting the powder charge properly for the lob trajectory and cutting the fuse so that
the shell exploded at the correct instant were skills not always available.
Mortar vessels were quite rare. It has been suggested that the French
colonel invented the exploding cannon shell in order to cause more
casualties in naval battles. This might well be true. A typical major
naval battle during the Napoleonic Wars would rarely have as many as
a thousand men killed while a land battle might cost more than ten
thousand. In fact, in the two major British victories in those Wars,
Waterloo and Trafalgar, of the 67,000 men engaged on land, 19,000
(or 28%) of the British side were killed or wounded and of 18,000 at
sea, 1,763 (or 10%) were killed or wounded.[35] The exploding cannon
shell was, as one would expect, an extremely dangerous weapon against
wooden ships with highly flammable rigging and sails. One response to
this danger was to attach armor plating to the sides of the hull,
although it would provide no protection to rigging or sails. In 1859 the
French navy launched its first and perhaps the world's first ironclad
with iron placed over the wooden structure of the ship (the *Gloire*).
The British responded in the next year with HMS *Warrior* with a com-

---

35. Keegan, *The Price of Admiralty*, p. 85 (1989).

plete iron hull over a wooden frame.[36] Fortunately for historians and those who enjoy beautiful ships, *Warrior* has been restored and can be seen at Portsmouth, England.

In the meantime steam had been applied to ship propulsion, first to small vessels, then as auxiliary to sail for larger ships and finally as the principal motive force for most naval vessels. After about 1800, the paddle wheel system was the initial choice as the way to transfer the power of steam to the water. But at least for naval purposes the paddle wheels were quite vulnerable to gunfire and shortly screw propellers replaced the paddle wheels. Somewhat counterintuitively, screw propellers turned out to be more effective as proved by the Royal Navy test described in Chapter 3.

These two trends, steam and armor, combined dramatically in 1862 in the American Civil War battle on Hampton Roads between the very innovative steam driven ship, *Monitor*, an iron ship with a single movable two-gun turret and *Virginia*, an ironclad wooden steamship with more or less conventionally placed armament In both cases the absence of sails and masts made the fields of fire of the guns more effective. While the battle was pretty much a draw, it did establish that either ship was more than a match for traditional wooden vessels. Thereafter, for the rest of the nineteenth century the major navies of the world devoted themselves to building steam-powered ships of iron and later steel with progressively larger guns, of longer range, breech-loading and firing explosive and armor-piercing shells. Gradually, as steam power made rigging, sails and most masts unnecessary, heavy armament was moved from inside the hull to the decks to obtain a wider field of fire. Obviously, the installation of rotating turrets beginning with *Monitor*

---

36. Colonel Paixhans' invention did not have the long-term effect that had been anticipated. The result of the continuous "thrust and parry" of military developments tends, over time, to blunt the effects of innovation. For example, in the Battle of Midway in 1942, the decisive naval battle of the Pacific War fought between two large, well equipped and modern fleets, the combined fatalities of the combatants were fewer than those of the Battle of Trafalgar which was fought entirely with solid shot. Keegan, *supra*, p. 211.

was one way to go but in the latter part of the nineteenth century many battleships had their heavy guns in barbettes, rather like non-moving gun turrets without roofs. This is not as reckless as it might seem. Since large caliber naval guns in those days did not have either long range or high trajectory, most rounds hit the sides of ships and thus the sides of the barbettes. A roofed turret only became necessary later when plunging fire become possible with longer range elevated guns firing from a distance, as was the case by the end of the century.

To the extent that there was a naval arms race in the 1880s and 1890s, Britain's Royal Navy won it. While it was usual at the time to rank navies on the basis of battleships in service, gross numbers of various categories of ships was not the only criterion of success in the race. Great Britain needed a much larger navy than any other country if for no other reason that it had far more territory to protect and that territory was spread across the whole world. Thus at any one point in space or time only a fraction of the Royal Navy could be present. At this time, however, even small fractions of the Royal Navy were exceedingly powerful.

Another factor beyond the number and areas of operation of battleships was especially important in the latter part of the nineteenth and early twentieth centuries. The penetrating power of the big guns and the resistance of the armor to them was a seesawing contest. The guns were increasing in size[37] and muzzle velocity, both factors which should have increased the penetrating power of the shells. The distance

---

37. Measured by the diameter of the bore in inches or millimeters for this purpose. While small arms, machine guns and other automatic weapons are usually described in caliber, meaning the diameter of the inside of the barrel (*e.g.*, .50 caliber for a one-half inch machine gun barrel, 20mm caliber for an eight-tenths of an inch cannon barrel), the description of larger naval weapons uses caliber to denote the ratio of barrel length to barrel diameter. Thus, the versatile primary and secondary battery cannon used by the United States Navy in World War II was described as 5-inch 38 caliber, meaning that the length of the barrel was 38 times the 5 inch bore, and was by virtue of the longer barrel a more effective weapon than its predecessor, the 5-inch 25 caliber.

which the shells had to travel to their target also was of great importance since shells slow down during the course of their flight, although not necessarily at a linear rate.

The competition was not just to make bigger guns and thicker armor plate. It was to make the guns capable of penetrating the armor with their shells intact so that the explosion, or most of it, would occur inside the target's armor belt. The idea of a non-exploding sabot round either had not arisen or it was felt that such a round would not be particularly effective given the large spaces in the interior of a battleship. Likewise, enhancing the armor protection was not merely making thicker plate but making more effective armor. To that end, initially when armor meant iron, thickness improved protection. As sandwiches of materials, different metals (steel and iron chiefly) and different types of the same metal, all designed to resist the impact of heavy shells and the explosions of the high explosive charges[38] within them, the comparisons became more difficult.

Of the rest of the world's potentially great navies, France and Spain were clearly in decline toward the end of the nineteenth century, Italy although having some great naval architects could not afford much of a navy, Austria-Hungary had relatively little interest in naval matters given the size and location of its coastline, Russia suffered seriously from having three seacoasts so separated that three separate fleets were necessary (Baltic Fleet, Black Sea Fleet, Far East Fleet), Japan was just getting started in the naval construction business with Britain's help, Germany was starting to build a serious fleet, and the United States was finally waking out of the military and naval stupor it entered after the Civil War

---

38.   Contrary to the belief of many, "high explosive" is not redundant. The term was coined to distinguish an explosion that was both powerful and sudden. The distinction was initially to gunpowder, a low explosive, and later to other propelling charges as contrasted to explosive charges. The difference is sometimes described as low explosives are heaving charges while high explosives are cracking charges.

At approximately the same time as the first powered flight, a revolution in battleship design was brewing. By the end of the nineteenth century, the battleship had evolved into an armored vessel armed with a variety of weapons, some in turrets with a wide traverse of fire, others in casements with a much more limited traverse. The latter were almost a carryover of the type of cannon mounts used by Nelson a hundred years earlier. As the main battery became made of larger and heavier weapons, a ship could carry fewer of them and each had to cover more area. While shipboard turrets date from *Monitor* in the American Civil War, turrets gained acceptance only slowly but steadily. Toward the end of the century, battleships carried a few of the largest guns in turrets and a number of smaller caliber weapons in a variety of mounts and for a variety of purposes. Not surprisingly, the smaller guns tended to have more limited range and less explosive power in their shells. Between the Civil War and 1900 considerable thought and theory was applied to naval gunnery, and several actual naval engagements in the Spanish-American and Russo-Japanese Wars had provided some practical experiences. As a result of these experiences, a few ship designers suggested that better battleships would result if there were more of the larger guns, turret mounted and all of the same caliber. At least three countries started down that route, Great Britain, Italy and United States. They proceeded at very different paces. Because the Royal Navy was at the time the world's largest and most powerful, it had the most to lose if a revolutionary development made a large portion of its fleet obsolete. It had fiercely resisted incorporating such modern and potentially obsoleting weapons as the self-propelled torpedo and the submarine for just that reason, although perfect foresight would have seen that both would imperil the primacy of the battleship. In spite of the close association between torpedoes and submarines during the twentieth century, the torpedo, meaning the self-propelled device not the mine of Civil War days, was designed initially to be used by surface ships against surface ships.[39] One of its virtues was that it would explode directly against the hull of a ship, increasing the effect of the explosion substantially. The second was that it could be set to explode at any desired depth. Third, it could be fired from some distance from its target. The submarine, although a much older invention, was only becoming practical by 1900, but the Royal Navy was not much interested in it either. While one could quite easily make a similar self-obsoleting argument against a new type of battle-ship, nonetheless Britain moved very rapidly to the forefront in this matter. It was surprising that the Royal Navy acted with such alacrity to build *Dreadnought* in 1906.

There has always been a dispute about who should have the brag-ging rights for having thought up the idea of the "all big gun" battle-ship. A famed Italian naval ship designer wrote an article proposing such a ship in 1903 (with twelve 12" guns) although the Italian gov-ernment had declined to spend the money necessary to act upon his idea. It is sometimes said that Japan might have a claim to priority in time, but the Japanese ships involved seem to bear a greater resem-blance to the last of the pre-dreadnought ships of the Royal Navy than to the subsequent ones. The United States claim is based on a Congres-sional appropriation of 1904 for a class of two battleships (USS *South Carolina*—BB-26 and USS *Michigan*—BB-27) with eight 12" guns, although they were not authorized until 1905 nor started until 1906. Even then the United States Navy was in no rush, for it was not involved in a fleet building competition as Britain was with Germany; the class was not completed until 1910.

Thus Great Britain led the parade to salt water in 1906, being force-fully propelled in that direction by the First Sea Lord, Admiral of the

---

39. In fact even battleships were sometimes equipped with torpedo tubes. These were not generally the aimable deck batteries of destroyers and cruisers but rather were underwater tubes firing broadside or, in a few early instances, astern. There is still a question as to whether HMS *Rodney* hit *Bismarck* with a torpedo in their 1941 battle.

Fleet Sir John Fisher.[40] Fisher believed in the all big gun battleship and supervised construction of the Royal Navy's first. Her name was HMS *Dreadnought*, a name that had been used in the Royal Navy for centuries. She was such a departure from prior ship designs that she would give her name not just to direct copies (as would normally be the case for the first ship in a class) but to all subsequent battleships. In fact because there were no exact copies of *Dreadnought*, there was not even a Dreadnought class although there were a great many dreadnoughts.

She was built in about a year, an extraordinarily short period given the usual battleship construction period of from two to three plus years. Part of the reason was that Fisher was pushing so hard, part was because eight of her ten 12" guns were diverted from other ships then under construction (the lead time for main battery guns was usually the longest for any battleship component), and part was that she was to be turbine powered. She was the first heavy naval ship in the world to have a turbine drive. She was also the first to have a single gunnery director for all of the turrets. Surprisingly, *Dreadnought* was not substantially more expensive than the battleships that preceded her; the difference seemed to be between 10% and 25% depending on the source and time of calculation.

As the ranges of naval battles had increased, it had become harder to hit a target. In modern sea battles, very few shells hit target vessels. Even as late as the Battle of Jutland (1916) between two powerful, modern, well-trained and well-equipped fleets of dreadnoughts, not more than 3% of the heavy shells hit their targets (and many of those did not explode). In the earlier Battle of Santiago fought at close range and under ideal conditions, at least for the Americans, only about 1 1/2% of their shells hit. The reason was largely the difficulty of finding

---

40.   First Sea Lord is a title held by a senior admiral in the Royal Navy. It is roughly the equivalent of the Chief of Naval Operations in the United States Navy and should not be confused with the First Lord of the Admiralty, a civilian political position akin to the United States Secretary of the Navy when that was a cabinet post prior to the reorganization of the defense departments in 1947.

the range of the target. Of the two components of aiming, range and azimuth, range is far harder to find. The target is most likely to be sailing more or less parallel to its enemy, presenting a target that is perhaps 100 feet wide The optical range finders would have had trouble coming that close given sighting errors and visibility problems. But added to these uncertainties are such additional factors as temperature of the guns, temperature of the powder, barometric pressure, winds, roll and pitch of the firing vessel, changes in the course of target during the flight of the shell, etc. It was calculated during World War II that variable factors in shell flight and target motion were so great that a German 88mm antiaircraft gun firing at a B-17 at 25,000 feet would not hit the B-17 even if the pilot tried to fly the plane into its path. So too with naval gunnery.

The solution to the high altitude antiaircraft problem was not to fire at particular planes but rather to set up a barrage of shells in a given area. With respect to naval gunnery the approach had to be somewhat different. The best range was obtained from the rangefinders, with the gunner making what adjustments seemed reasonable and firing. The splashes of the falling shells were observed and the range was adjusted. The objective was to have one set of splashes on one side of the target and the next on the other. The following shells would be directed between the two. This does not mean they would hit for among other factors the gap between the two preceding sets of splashes might have been very large or not centered on the target, both ships would have moved in the meantime and the target may have been following an old defensive tactic—turning toward the last fall of shot to throw the off the adjustments to range made in the foregoing fashion.

All range adjustments were complicated by the fact that individual gun and turrets had been separately controlled and different caliber guns were firing at the same time. Thus a spotter might assume that a shell splash belonged to his gun when it belonged to another similar gun, or worse belonged to a different caliber gun with a different trajectory, or worst was in fact fired at ship A but the spotter assumed it were aimed at ship B and adjusted accordingly. While one might suppose that the explosions of different caliber would be distinguishable, one of the lessons of the Battle of Tsushima was that this was not so. By combining a number of identical big guns and firing broadsides (all guns which could bear on the target at once), correcting the aim of all was easier. When the aim was correct, a number of big gun shells would fall in a rather compact area, increasing the probability of actual hits. Additionally, having all guns of the same caliber would mean that all of the ammunition and gun parts would be interchangeable. Fisher meant the "all big gun" phase literally. He rejected the idea of secondary batteries (4" to 6" guns) because he was sure that the battleship would engage any enemy far away from the ship and thus close-in weapons would not be necessary to deal with destroyers and torpedo boats before they could approach into torpedo range. This was a particularly curious position because the Royal Navy had felt that the biggest danger it faced from the French Navy was from torpedo boat attacks in the Atlantic and Mediterranean.[41] While he was aware of some risk from torpedo boats, he evidently felt that the even lighter armament (1 1/2"–3") which was to be installed would suffice if any torpedo boats survived the 12" gunfire of the main batteries. But torpedo boats were becoming larger as were destroyers, requiring more powerful guns to deal with them. Furthermore, hitting such small and elusive targets was a job better handled by more rapid-firing weapons than the main batteries. A secondary battery was really a necessity. In fact a number of classes of subsequent British dreadnoughts continued to be built without any secondary armament but, beginning with the 1914 completion of the first of the Iron Duke class of battleships and with HMS *Tiger* (battlecruiser), also competed in 1914, secondary batteries of 6" guns became standard as clearer heads prevailed in the Admiralty. The Germans, on the other hand, put secondary armament (5.9" guns) on all

---

41.   Massie, *Dreadnought*, pp. 426, 451 (1991).

of their dreadnoughts, battleship or battlecruiser, beginning with their first completions in 1909.[42]

A battlecruiser was a new type of naval vessel just coming into service in the early twentieth century. She was a heavy vessel about the size of a battleship with battleship type armament (11" to 15" main battery) but with considerably thinner armor. The weight savings were translated into speed. The class was not generally successful because the battlecruisers were expected to fight with battleships on occasion but their armor was seriously inadequate to resist the shells of either battleships or battlecruisers. A few battlecruisers appeared in World War II. The Royal Navy had two conventional vessels, HMS *Repulse* and HMS *Renown,* and one very large one, HMS *Hood,* all started as World War I construction. Germany built two battlecruisers for World War II service, *Scharnhorst* and *Gneisenau.* The United States for some reason started two battlecruisers, *Lexington* and *Saratoga,* at the end of World War I but before they were finished they were effectively outlawed by the Naval Treaty signed in 1922 and had to be converted to aircraft carriers, to the great if inadvertent advantage of the United States Navy in World War II. The United States came back into the game in World War II with a class of very heavy cruisers armed with 12" guns and 27,500 ton displacement (the Alaska class).[43]

Some nations considered building battlecruisers between the wars for different reasons. While they realized that battlecruisers could not really fight battleships, there were two reasons why that disadvantage might not be determinative. First, battlecruisers and battleships were in

---

42.   Massie, *supra,* p. 909.

43.   Morison, *History of United States Naval Operations in World War II,* vol. XV, p. 35 (1962) Since the usual main battery of heavy cruisers was 8" (a carryover from the naval treaties of the 1920s) and the usual displacement was from about 10,000 to 13,500 tons (a partial carryover), some considered members of the class to be more like battlecruisers even though their top speed was no greater than that of the most recent class of United States battleships. Compounding the confusion, in the United States Navy cruisers were then named for cities, battleships for states and the Alaska class for "territories and insular possessions."

the same category of ships regulated by various treaties, and the permitted number of ships in the category was calculated by aggregate tonnage rather than by the number of hulls. It was thus sometimes possible to fit an extra ship into the battleship category by using battlecruisers because they were lighter in armor and sometimes had one main gun turret less. Second, some far-thinking people realized that the extra speed of the battlecruiser would enable it to work better with aircraft carriers which had to have great speed to assist in launching and recovering aircraft.

In the historical broadside actions for hundreds of years the vast majority of shipboard guns were mounted on the sides of ships. Traversing those guns was a difficult matter and the range of possible traverse was very limited. Even after turrets came into use, broadside actions were still considered desirable both because more weapons could be brought into play along the side of the ship than over the bow or stern and because range was a harder component of aim to find than azimuth. Accordingly, a ship firing broadsides had more available fire power and was harder to hit. It was for this reason that "crossing the T" was often a highly desired tactic in fleet engagements, both under sail and steam. It was, however, not always successful nor even desired.

One fleet crosses the T when it places a line of ships in front of, and at right angle to, an enemy column. In that event all of the broadside guns of the crossing fleet (which, except for the original dreadnought battery placement, meant all of the heavy guns) would bear on the leading ships of the enemy, which could only counter with the forward-firing guns of the first few ships. The situation was made even for difficult for the enemy by the fact that the usual way to escape was for each ship in turn to reverse course 180 degrees away under heavy fire, exposing the next ship to the same battering. The German battle fleet at the Battle of Jutland (misnamed the High Seas Fleet—it never reached the high seas) did escape from two T crossings only by causing all of its ships to execute simultaneous 180 degree turns away, risking collision and confusion but succeeding.

At Trafalgar, Nelson's basic tactic was to let the French/Spanish fleet cross both T's of the Royal Navy formation. By doing so he accepted the fact that his ships would be fired upon by the broadsides of the French/Spanish fleet for many minutes (because of light winds) before his ships could reply effectively. He felt, obviously correctly, that his ships could stand that fire and would, at the end of the maneuver, break enemy's line of battle into three pieces which the Royal Navy could and did then destroy.

Fisher seemed of the same mind as Nelson. He favored his ships charging at the enemy ships rather than pounding away at them in a parallel line. Thus, he wanted *Dreadnought* to have heavy firepower forward. British naval thought was to the effect that superimposed turrets of large caliber (one turret behind the other but raised so as to have a clear field of fire) would not work because the blasts of the guns from one turret would interfere with the other. This view was quite wrong, as was established when *Michigan* and *South Carolina*, begun at about the same time as *Dreadnought*, went into service in 1910 with superimposed turrets fore and aft working perfectly well, although the British would have had to make a small change in their usual turret design in order to superimpose without problems.

Nevertheless, to avoid superimposition and yet to provide maximum firepower forward, *Dreadnought* had an unusual turret layout. She had one turret in the bow, turrets on each side of the bridge (wing turrets) one amid ship and one in the stern, all containing two 12" guns. All except the bow turret were on the same main deck level. The wing turrets could fire to their respective sides of the ship, the bow turret forward, the stern turret aft and the mid-ship turret only to one side or the other. Nominally therefore, six guns could be fired forward, eight abeam and six aft, still favoring broadside firing.

All is not as it seems from the foregoing. Careful analysis of the firing arcs of the five turrets shows rather a different result. The bow turret could fire from about 26 degrees abaft the port beam to about 50 degrees abaft the starboard beam but with an opening in the firing arc from dead ahead to 30 degrees off the port bow. The wing turrets fired in a 183 degrees continuous arc from 1 degree across the bow to 2 degrees across the stern. The mid-ship turret covered roughly 70 degrees on one side and 120 degrees on the other side and the stern turret 130 degrees to port and 150 degrees to starboard.[44]

Thus in broadside fire eight guns were available, all three center line turrets plus the near wing turret. For forward fire, if the target were dead ahead to 1 degree to starboard, three turrets (bow and both wings) could fire. If the target were more than 1 degree off the starboard bow, only two turrets (bow and starboard wing) were available. If it were dead ahead to 1 degree off the port bow, only both wings would bear and if the target were more than 1 degree but less than 30 degrees off the port bow, only the port wing turret was available. After 30 degrees to port both the bow and port wing turrets were available. Astern, both wing turrets and the stern turret were available ±2 degrees from dead astern; if the target were further to the side, only the stern and appropriate wing turret were available.

There may be some question about the port side gap in the firing arc of the bow turret. John Roberts, in his very detailed book, *The Battleship Dreadnought* (rev. ed. 2001), does not mention the gap in his textual description of the fields of fire of the big guns but surprisingly he does not include a diagram of the firing arcs despite the minutia otherwise contained in the book. He does, however, raise another point about the arcs that produces a similar effect. He points out that both of the wing turrets, despite their theoretical firing arcs right forward and right aft, could not have been fired over the full range of their arcs because the blasts would have damaged the superstructure at each end of the arc. If the effective arcs were reduced only 5 degrees from dead ahead or dead astern plus losing the 1 degree overlap on the bow and the 2 degrees overlap on the stern of each wing turret, there would be a significant effect on the firepower both forward and aft. In either the

---

44.   Massie, *op. cit. supra*, p. 473.

Massie or the Roberts view, the relative firepower fore, aft and broad-side was similar.

What all this establishes is that, on Massie's facts, *Dreadnought's* best field of fire was broadside, her next best was astern and her worst was forward, the direction Fisher favored. While, again using Massie's facts, there was a 1 degree arc forward in which three turrets could fire, holding a target in that arc would be very difficult, especially since that target would presumably be firing broadside and thus moving at an angle to *Dreadnought*, producing rapid azimuth changes. Fisher described broadside fire as "peculiarly stupid." He felt that "to delay your pursuit by turning even one atom from your straight course on to a flying enemy is to me being the acme of an ass."[45] Yet he would have known, of course, that unless the enemy is stopped dead in the water or sailing directly away from or toward you, the fastest way to reach him, if both ships are steering straight courses and each is holding a constant speed, is to keep the enemy at a constant bearing. That bearing would be off either the port or starboard bow. For example, if both ships are making the same speed and the courses are 90 degrees apart, the target should be held at 45 degrees off the bow. To the extent that the courses are not 90 degrees apart or the speeds are different, the angle off the bow will be different and would have a good chance of being within the 30 degrees to 1 degree arc on the port bow where only one turret would bear.

Using what Roberts might be implying although he does not say so and assuming again only a 6 degree loss forward and 7 degrees aft of each of the wing turrets, for targets from 5 degrees on the port bow to 5 degrees to starboard, only the bow turret would bear. For targets on other forward bearings, two turrets, bow and the appropriate wing would bear. For broadside fire four turrets, bow, wing, mid ship and stern (or A, P or Q, X and Y in Royal Navy terminology) would bear. Astern only the stern turret could fire at targets within 5 degrees of

---

45.   Massie, *op. cit. supra*, p. 472.

dead astern. Outside of that arc the stern turret and one wing could fire until the target reached an angle where the full broadside was available.

Fisher demanded speed—at least 21 knots. At this time battleships typically had top speeds of 18 knots and that number is seriously misleading. Battleships had as power plants reciprocating steam engines, rather like automobile engines except that instead of a gas/air mixture exploding in cylinders to generate power, high pressure steam was inserted for the same purpose. These steam engines were very complicated and very noisy. They were also very inefficient in that the mass of the pistons had to be reversed several times a second at high speed. Not only was the top speed limited by this type of engine but also by the time during which that speed could be maintained. The latter limiting factor was not the availability of steam but rather the wear and tear on the engines. Massie says that a four-hour top speed run by a pre-dreadnought battleship with reciprocating steam engines could require ten days of yard work to repair and readjust the engines.[46]

In 1897, at the annual fleet review of the Royal Navy, a strange small vessel called *Turbinia* traveled at high speed (30+ knots) through the fleet, producing some interest but probably more indignation at her effrontery. She was, as her name suggests, turbine powered and was a private venture of Charles Parsons, a turbine manufacturer. A turbine engine uses a flow of gas (high pressure steam or combustion products) to turn a bladed wheel which in turn applies that power to a ship's propellers. It is, in a sense, a paddlewheel working backwards. The power supply is smoother and the engine is much simpler and quieter.

A similar contrast occurred in the automobile industry some years ago when the Wankel rotary engine appeared. While not a turbine engine, its advantages were those of a turbine and Mazda used it for a while in some of its cars. Technical problems prevented it from supplanting the reciprocating engine for general automobile use but, had the same time, money and energy gone into improving the Wankel as

46.   Massie, *op. cit. supra*, p. 474.

had gone into the conventional internal combustion engine, the Wankel might be powering most cars and trucks today.

The ships' turbine proved to be more successful than the Wankel and had previously been tried on a few smaller vessels. To get the speed needed, Fisher used turbines in *Dreadnought* with great success. After *Dreadnought*, all British and most other countries' heavy ships used them exclusively. One notable exception was the United States Navy. While it was very quick to adopt the "all big gun" battleship (or perhaps even to initiate it), its first two dreadnoughts were conservative in their power plants, keeping reciprocating steam engines. As late as the beginning of the First World War the United States was still building some reciprocating engine battleships (USS *New York*—BB34, USS *Texas*—BB35 and USS *Oklahoma*—BB37), although interspersed with turbine powered ones (beginning with USS *North Dakota*—BB29). While *Oklahoma* was destroyed at Pearl Harbor, *Texas* and *New York* served throughout the Second World War with their reciprocating engines. Turbine engines improved over time and, when geared turbines appeared, their cruising fuel economy began to match that of the reciprocating engines. Steam turbine engines endured throughout the twentieth century until being replaced in some applications by gas turbines with much shorter starting times.

Curiously, some recent research has suggested a different view of Fisher's original objective. It is thought that he may have preferred dreadnought type battlecruisers to battleships, feeling that their higher speed (by four knots or so over even the proposed turbine-driven battleships) would enable them to control the action between the battle fleets.[47] But battlecruisers and battleships tended to have the same main batteries and thus the same range and hitting power. So at any given range, all else being equal, each would have the same chances of making hits on their opposite numbers. Those hits would be much more damaging to the battlecruisers. There seem to be three ways that Fisher could have thought that his battlecruisers could have an advan-

---

47.   Padfield, *Battleship*, p. xiii (2000).

tage in a fight with battleships. First, they might have a better fire-control system, allowing them to make a significantly higher percentage of hits; second, they might use their higher speed to maneuver into positions in which they could apply a higher portion of their firepower; and third because they were somewhat cheaper to build more of them would be available (the issue of compliance with international treaties limiting the size of the navies of major powers did not arise until after World War I). Dealing with these "advantages" in order, if a better fire-control system were available to battlecruisers why would it not be installed in battleships as well, negating any advantage? Secondly, while having a positional advantage is very important, getting it is difficult for many reasons, speed being only one factor. The Battle of Jutland is a good example of the difficulty of knowing the position of the enemy in an era without radar or aerial reconnaissance. Finally, the cost differential is probably not significant, particularly when operating costs are factored in.

It is also difficult to reconcile Fisher's views that he wanted a lot of firepower forward on his heavy ships so it could be brought to bear on his enemy while his ships were charging in to close the range with the new theory that he really wanted battlecruisers in preference to battleships so that their greater speed could be used to control the battle by, among other things, moving out of range by virtue of their superior speed. Perhaps Fisher was just being expedient and asserting what he thought was useful to his positions from time to time, or perhaps some of the positions did not have much merit over time. Somewhat surprisingly, the Royal Navy did build a small class of post World War I battleships with all the heaviest guns forward, HMS *Nelson* and HMS *Rodney*. Both carried the heaviest guns the Royal Navy mounted in the twentieth century, three 16" three-gun turrets, all on the centerline. The middle turret was superimposed over the forward and aft turrets; the third turret was back on the main deck but still forward of the bridge. The result was that only two turrets would bear on targets ahead and none would bear much abaft the beam. On Fisher's con-

stant bearing closing course, however, these would have been deadly
vessels as happened in the pursuit of *Bismarck* by *Rodney*. This class of
ships had a strange silhouette due to the turret arrangement. One of
the few amusing matters to come out of the first and last cruise of *Bismarck* was when a senior Royal Navy officer mistook *Rodney* for *Bismarck*, an impossible gaffe as the officer ruefully admitted in telling the
story on himself.

The three forward turret down, up, down arrangement was unique
to the Nelson class for battleships. A similar arrangement was sometimes used for cruisers but they usually had two stern turrets as well,
one superimposed. The cruiser arrangement avoided one potentially
serious problem of the Nelson class; with the main turrets that close
together, the whole heavy firepower of the ship could be wiped out by
one hit.

The Battle of Jutland is often cited as proof of the folly of exposing
battlecruisers to battleship fire, or battlecruiser fire for that matter,
since three modern Royal Navy battlecruisers were blown up by maga-
zine explosions caused by enemy shells. While the conventional explanation is that the Royal Navy had neither good enough protection
from enemy shell explosions that flashed down to the magazines nor,
in their zeal to speed up their rate of fire, the discipline to keep what
protections there were in use, that explanation may not be the whole
story. It is felt in some quarters that the instability of the British cordite
propellant was the principal cause of these sinkings. Padfield describes
it best as German propellants burned when hit, destroying turrets,
British propellants exploded when hit, destroying ships.[48] A fairer
example of the inherent vulnerability of battlecruisers would be the loss
of *Hood* in World War II, almost certainly the result of plunging fire
by a heavy shell penetrating the lightly armored deck and reaching an
ammunition storage area.

Fisher was opposed to gun turrets on the main deck. He felt
strongly that they should be mounted higher to avoid wave interfer-

---

48. Padfield, *op. cit. supra*, p. 241.

ence in high seas. Rather surprisingly, he lost this argument because of stability concerns but *Dreadnought's* bow turret, the most exposed to wave action, was raised. The other four turrets were placed at main deck level.

If one reviewed Fisher's contribution to larger ship design at the time of World War II, one would see that the all-big-gun concept as Fisher wanted it had been watered down extensively. All battleships (and battlecruisers for that matter) had multiple sized guns, although many were for anti-aircraft protection which Fisher could not reasonably have anticipated. All battleships had at least some of their main battery turrets based on the main deck. All battleships used superimposed turrets. Almost all battleships, while essentially still favoring the broadside volume of shots, could fire heavily forward well, as Fisher said he preferred but contrary to what *Dreadnought* could actually do. The standard late-model United States battleships had three main battery turrets, two forward (6 guns) and one aft (3 guns). The Royal Navy's King George V class also had two forward (6 guns) and one aft (4 guns); a preceding class (*Nelson* and *Rodney*) had three forward (9 guns) and none aft. The French Navy had several battleships (Richelieu class) with two turrets forward (8 guns) and none aft. The Japanese "super" battleships of the Yamato class had two turrets forward (6 guns) and one aft (3 guns). Only the Germans held out with two turrets forward (4 guns) and two aft (4 guns) in *Bismarck* and *Tirpitz*, although even they mounted two turrets forward and one aft on their battlecruisers. The path to this result was not, however, a smooth one. Both the Royal Navy and the United States Navy went through periods of midship turrets, one, two and in at least one instance three (HMS *Agincourt*, for which there is a separate explanation) only usable in broadside actions.

While the *Dreadnought* had some teething problems on her first cruise and had the weaknesses mentioned above, it was clear to the whole world that she made all previous battleships obsolete. Her main battery, her fire control system and her speed were all in a class of their own. If any nation planned to take on the Royal Navy, that nation would have to build a whole new battle fleet of similar ships.

At this one launching Great Britain threw away its huge lead over Germany in battleships—the score was now only one to zero in favor of the Royal Navy whereas it had been 62 to 8 in predreadnought battleships in 1897. By the then best estimates of British and German production, the ratio of dreadnoughts in 1912 would be between 20 to 17 in favor of the British and 21 to 20 in favor of the Germans.[49] Britain and Germany now started a new naval race. While it was one which Great Britain could perhaps win, the Germans felt that, even if they could not quite equal the British by the time war began, they might be able over time to pick off a few British capital ships and reduce the odds to about even or better. Then they might be able to hazard a general fleet engagement if the war required it.

It is difficult to appreciate, at this far removed time, the political and economic factors that went into the dreadnought race. Governmental budgets were substantially altered by decisions to build or not to build several additional dreadnoughts in a particular year. Even after the fleets were set at the beginning of World War I, an admiral who risked several dreadnoughts to enemy action was not only risking the ships and the men in them, he was risking large amounts of national assets as well as his country's strategic position in the war. The pressure on the decision makers, whether naval or civilian, must have been tremendous and must have pushed hard in the direction of conservatism. Conservatism can be a virtue or a fatal flaw in war. Knowing whether it is the correct course at any given time is the mark of a great commander. The underdog in a particular tactical situation is unlikely to prevail through conservative actions; he must take risks by making unusual moves. On the other hand the more powerful force may very well stick to the conventional approach and achieve the desired result.

It is curious how eager countries were to go to war in August 1914. On top of the generalized risks inherent in war, the naval situation was,

---

49.   Keegan, *op. cit. supra*, p. 105.

at least in terms of objective hardware (quantity, not necessarily quality) and numbers of seamen, in a delicate balance, and the long-planned German army moves against Belgium and France were orchestrated down to the day with even a slight delay likely to be fatal to the Schlieffen Plan. The risks to the Germans of starting to go to war seem to have been, in hindsight of course, too great to have been taken rationally. Once started, however, stopping would probably have been both psychologically and logistically impossible.

Now, at long last, appears the reason why the building of *Dreadnought* has been of great importance to the twentieth century and perhaps longer. Had Great Britain not built *Dreadnought* when it did thereby allowing Germany to come near to catching up with the Royal Navy, Germany would have been undoubtedly been more reluctant to become involved in the escalation of tensions resulting from the death of the Archduke Ferdinand in June 1914. Notwithstanding the pretexts based on national honor and morality, the domino effect of the various alliances started by his death might well have been averted by diplomacy if the Germans had wanted to negotiate or had feared the consequences of a rush to war. As it was Germany was anxious for war as an opportunity for it to take its "rightful place" as the dominant nation in Europe and confident its view of "social Darwinism," suggesting domination by the Germanic peoples through natural selection, would prevail. But without either control of or a standoff in the seas around Germany, Germany had serious problems that might have tempered the Kaiser's ambitions, at least temporarily. Of course the dreadnought idea was not solely a British creation; the United States had designed a class of dreadnoughts before *Dreadnought* was started and that class would have been built in any case. But it would have been at least four years until those ships were commissioned and probably several more years until they were proven to such a degree that the Germans and British would decide to join in a building race. Even then, while the American gun placement arrangement was superior to *Dreadnought*'s plan, the American power plant was inferior. Who knows what might have happened if the six or more years had elapsed? Whatever happened it could hardly have been worse than World War I and the resulting World War II. Perhaps Russia would not have had its 1917 revolution without the pressures on it caused by World War I and the serious losses it suffered although there is certainly a valid counter argument that the Russian government was glad to have the German and Austrian armies massing on its frontiers so that Russia could try to execute the old political strategy of using a foreign war to distract and unify the country in the face of serious internal difficulties. Without World War I, World War II was much less likely as was the Cold War. In those events the twentieth century world would have been much different and the odds certainly favored "better," recognizing that some major twentieth century events would have been necessary to address the strains created by the events of the nineteenth century.

In saying that there might not have been a World War II under these circumstances I do not mean to suggest that there would not have been a major war in the Pacific in mid-century. The factors that might have reduced the eagerness of the European powers to rush to war in 1914 would have had no relevance to Japanese ambitions. Having defeated the Russians soundly in the Russo-Japanese War of 1904-05, they went on to occupy Korea in 1910 and various parts of Manchuria, China, Sakhalin and even a bit of Siberia in the same time frame. After World War I the Japanese continued to expand, generally to obtain sorely needed natural resources (all of Manchuria-1931) and food (more of China-1930s). As time went on they became more aggressive, moving well into Southeast Asia in 1940. At that point the United States declared an embargo on sales of certain commodities from the United States to Japan until Japan withdrew from Southeast Asia. The most critical products under embargo were fuels. Japan's desperate need for aircraft and naval fuel to continue its expansion led to its attack on Pearl Harbor in 1941 and its invasions of resource-rich Dutch East Indies and Malaya and the intervening Philippine Islands.

These attacks would probably have occurred in any event and in fact in this scenario the United States would likely not have been as well prepared to battle the Japanese in the Pacific because there would not have been a European war raging for two years previously to goad America into some degree of preparation. Nevertheless, the United States would probably had no great difficulty in defeating the Japanese because without a European war to fight simultaneously it could have devoted all of its resources to the Pacific rather than the 15 to 30 per cent as in fact happened

Even had Germany not been dissuaded from war as a result of the huge disparity with Britain in the matter of predreadnought battleships (if *Dreadnought* had not been built and its innovations postponed until the American dreadnoughts had been completed and proved effective), the German nation might well have been defeated in World War I more easily by an invasion of northwestern Germany under the cover of the overwhelming power of the Royal Navy. There would have been no prospect of serious opposition from the predreadnought German navy. While this scenario would not have produced as dramatic an alteration in history as a total absence of World War I, it would also have changed the rest of the twentieth century significantly, and again probably for the better.

The battleship technology driven by the Dreadnought model through the rest of the battleships and battlecruisers ever built, unlike the other technologies discussed herein, did end as a practical matter in mid-century so its continuing effects were only as a result of the effects felt around the time of World War I but speculation about those effects certainly does allow and probably requires the conclusion that they were very profound in their influence on the future. The other three technologies did in fact continue to evolve rapidly and in most respects dramatically throughout the century and beyond.

The hero of the Dreadnought saga, like the technology involved, differs from those of the sections on automobiles, airplanes and radio. Admiral Fisher, like Albert Einstein, made no money from his idea; it was created and implemented as a part of his job as First Sea Lord. Many would say that his reward was the result of the Battle of Jutland.

To review that result we have to consider where matters stood at the beginning of World War I. Great Britain and Germany had been in a frantic naval arms race since the launching of *Dreadnought*. Germany was attempting to match the Royal Navy battleship for battleship and battlecruiser for battlecruiser, all of the dreadnought type. The attempt failed in the sense that at the beginning of the war the Royal Navy led by a count of 29 to 18, a considerable variance from the estimates of a decade earlier. The mere ship count is not necessarily the whole story. The German main battery guns were of somewhat smaller size although their ships may have been better in other respects—better range finders, a better quality of armor and better compartmentalization. The better range finders, all else being equal, should have resulted in more hits even though the German shells were somewhat smaller and thus less powerful. The better armor and compartmentalization contributed to keeping the German ships afloat even when heavily damaged.

After some inconclusive skirmishes early in the war the Royal Navy retired to its bases on the east coast of Great Britain and the Germans to their bases at Wilhelmshaven and Cuxhaven to await events. It was the German hope to be able to isolate small groups of British capital ships and attack them with much larger fleets. If this tactic were successful the Germans hoped to even the odds for a major fleet engagement. The attrition tactics did not work and the major engagement, the Battle of Jutland, occurred with both fleets at full strength although the Germans did not realize that at first.

The result of that battle was the subject of considerable debate for years. It was true that the losses of both ships and men were in Germany's favor. On the other hand the Germans retired to port to avoid further contact with the Royal Navy. In addition, many of the surviving German ships were seriously damaged and required lengthy repairs while almost all of the surviving British ships were ready for sea in a few days. It was felt by many neutral commentators that the strategic victory belonged to the British. Later events seemed to bear this out since the German High Seas Fleet never saw the high seas and at the end of the war was surrendered to the Royal Navy, was interned at the British naval base at Scapa Flow and scuttled itself in 1919. While one could argue that it was Fisher's fault that the Germans had been able to build a fleet that could challenge the Grand Fleet, once the decision on *Dreadnought* was made, Fisher had created a ship capable of defeating the Germans and had prepared the sailors, especially the officers, to sail and fight it. And in fact defeating the Germans was not required; all the British had to do was to avoid losing. This was not in the Nelson tradition and therefore not very satisfactory but it was enough. Admiral Jellicoe who, as the only person who "could have lost the war in an afternoon," did not.

As we have seen, the effective life of the Dreadnought "species" was about half a century. During that half century the battleships did fight as they were intended to, but not very much. Certainly the battleships did not do enough in action to warrant their consideration as a category of the importance of automobiles, aircraft, atomic energy or radio. The importance of dreadnoughts was their effect on important events during their brief existence and the consequences far into the future. A parallel situation, although as usual the time scales are vastly different, appears in natural evolution. During some portion of the Paleozoic Era (the Carboniferous and Permian Periods, the former sometimes being referred to as the Pennsylvanian and the Mississippian) and the Mesozoic Era (the Triassic, the Jurassic and the Cretaceous Periods) enormous quantities of dead vegetable matter were covered with sediment and over millions of years were converted to coal. Of somewhat less certain origin, other vegetable matter and possibly small animal matter, some probably dating back one billion years or more, was converted into petroleum. While most of the types of plants and animals involved have become extinct, the remains of those organisms have had a great impact on life on Earth in recent centuries even though the species involved were no longer around. Although the impact of a species usually appears to us through its descendents however remote, there are occasions where effects created by those organisms have had an influence on later times well beyond the effect of their descendents, if any.

7

Radio

I have left radio until last, not because it was the least important of the
five technologies but because it may have been, in some respects, the
most important. This may seem strange but read on, I may be able to
convince you.

Rapid communication is an objective of most governments and
businesses and is a desirable benefit to most societies. What "rapid"
means has varied dramatically over the past 5,000 years. For most of
that time, rapid meant about as fast as a sailing ship could sail with a
favorable wind or a man on foot or a horse could travel a short distance. It may surprise those who have not had the dubious pleasure of
long distance travel on horseback that the pace of a person walking and
a horse covering the same distance are very similar although the horse
could carry more baggage and its rider would be less tired than the person traveling on foot. Of course, all of the foregoing assumes that any
message to be transmitted rapidly would be oral. To the extent that the
message was in "hard copy," the problem was worse. Once writing had
been developed (over 5,000 years ago in Mesopotamia and possibly in
the Indus Valley), it would be likely that at least some messages would
be written, trading off the risk of interception for the certainty that the
message delivered was the message sent. In cultures that wrote on light
materials (paper, hides, birch bark) the difficulties were not great, but
consider the problems created if the messages were on clay tablets or
carved stones.

The land types of rapid travel could be improved in several ways.
First, the roads (or paths) traveled could be improved. Second, if a

vehicle were involved, it could be made faster and less prone to break-downs. Finally, instead of one man or one horse or one team of horses making the whole journey, relay stations could be established on a much used route at which horses and/or men could be stationed. If this were done, the horses could be driven at a much faster pace because they could be changed often. This was a very expensive procedure in terms of people, horses and general wear and tear on the equipment involved. It was done on many occasions, however, particularly on mail routes in Europe in the seventeenth and eighteenth centuries and stage coach routes in the United States in the eighteenth and nine-teenth centuries, although this relay concept may have been pioneered by the Mongol Empire in the time of Genghis Khan. Perhaps the most famous rapid communications route of this ilk was the Pony Express in the Western United States in mid-nineteenth century. By using fre-quent changes of fast horses and young, lightweight riders, mail could be delivered over the almost 2,000 miles from Saint Joseph, Missouri, at the western end of the eastern railroad line to Sacramento, Califor-nia, in about 10 days. It was expensive because it involved over 100 sta-tions, 80 riders and 400-500 horses, and was very short lived. A parallel telegraph line was opened 18 months after the Pony Express started with much faster, more reliable and cheaper service, driving the Pony Express into insolvency.

Some empires containing vast expanses to cover with instructions from the capital developed well organized but slow message delivery systems. The "neither snow nor rain nor heat nor dark of night" motto was not invented by the United States Postal Service or its predeces-sors; it was a contemporary description of the Persian Empire's messen-ger system of 2,500 years ago.[50] And the Inca empire of 600 years ago had very efficient systems of high altitude runners in the Andes. Never-theless this type of communication had a long and frustrating delay between sending orders or asking questions and receipt of confirmation

---

50. Herodotus, *Histories* 8:98.

of receipt of those orders or the answers to the questions. People lived with the delays because they knew no other way.

The Roman Empire, and the republic before it, developed a hard-surfaced road system which eventually reached a length of some 50,000 miles and ranged from the British Isles to North Africa and Asia Minor. The purpose of the network was initially military; it enabled the legions of Rome to travel rapidly, at least by the standards of those days, to deal with barbarian incursions or internal civil unrest. A side benefit of the road system was its ability to enable messages to and from far distant points to be sent with some degree of speed and reliability. The was particularly important because Rome had a highly centralized government under both the republic and much of the time of the empire. Although message delivery was made somewhat easier by the central location of the city of Rome, long delays and uncertain deliveries plagued efforts to control some of the outlying areas. When Rome finally gained control of the Mediterranean Sea by defeating both the Egyptians and the Carthaginians and by suppressing the widespread piracy, officials in Rome were delighted that they could now communicate by sea with many distant parts of the country in only a week or two with a high degree of reliability.

The same type of problems, although exacerbated by the greater distances, existed with later communications either to ships in the great oceans or to overseas destinations. Sailing ships seldom could make more than five knots and their courses could be very circuitous because of winds and currents. Furthermore many ships were lost, and thus any messages they were carrying were not delivered. It was common for many ship-borne messages to be sent in several copies in hopes that at least one would be delivered. The same duplication would probably have been necessary for the response. Any number of battles in the era of open seas naval engagements did not happen as planned because the necessary facts were not timely reported to the persons involved. Because of delays in communications that could last six months or more, captains and admirals in the great navies of the seventeenth through the mid-nineteenth centuries had extraordinary discretion to govern their own actions.

While there are innumerable examples of delayed communications affecting major events, one of the more notable was the failure of notice of the Peace of Ghent to be delivered to either the Americans or the British before the Battle of New Orleans was fought. The effects of this completely unnecessary battle, which was a disaster for the British, caused many Americans to believe that they had won the War of 1812 and probably made the American commander, Andrew Jackson, President of the United States. One could view that result in different ways, of course.

There had always been faster ways to convey simple meanings. If, for example, one wished to report major events, such as the birth of an heir to the throne or an invasion of the country, by pre-arrangement church bells could be rung at unusual times or on a clear night fires could be lit on hilltops. When these signals were heard or seen, they could be repeated and the message could cover hundreds of miles quite rapidly. In order for this type of one-way communication to work, weeks of effort would have been necessary to set up the system to be used and inform the recipients what the signal meant. Similar types of notice could be given at some distance in areas of high sunlight by reflecting mirrors, but again some pre-arrangement of the meaning of the message would have to have been given to all parties. Perhaps most famously, in North America and particularly in the western part, Indian smoke signals could be read at long distances in clear still air in daylight. Sending the signals from high ground increased the distance that could exist between sender and receiver.

Two of these types of communication, unusual ringing of church bells and controlled reflections of the sun, had not completely died out in the mid twentieth century. During World War II, church bells were to have been rung in Britain if the Germans invaded, an unnecessary precaution as it turned out, and small reflecting mirrors with instructions on how to direct the sun's rays were included in a number of survival kits issued to aviators. Of course, in this case the signal only delivered a message that someone was there, nothing more, although that was usually enough.

Things began to change rapidly toward the end of the eighteenth century. Use began to be made of the telegraph. Before thinking that there is confusion in the author's mind about when Mr. Morse did his work, note that he did not invent the term "telegraph." The word means simply writing at a distance and was properly used in the late eighteenth century to describe a new and effective way of transmitting certain types of messages rapidly. This system only worked on land and on clear days, but under the right conditions it worked very well. The telegraph system required a chain of buildings within sight of one another extending from the source of the message to the recipient. Each building had on top of it a mast to which there were attached wooden arms that could be moved by ropes from inside the building. The operators moved the arms from position to position to indicate letters, numbers or particular short messages, much like the semaphore system using hand-held flags which is still used for short range purposes, usually at sea. This type of telegraph system was particularly suited to send messages between fixed points with much message traffic, such as between naval headquarters in the capital and naval ports. It was so useful that enemies were known to raid coasts just to destroy the telegraphs which, because of the very complicated rope arrangements which controlled the signal arms, were difficult to rebuild quickly.

Ships at sea developed their own message systems to communicate with each other at relatively close range, initially by loosening the sheets on certain sails or by different flag hoists on different masts but eventually there evolved a system of quite sophisticated letter and number coded hoists that could convey considerable information. While the signals were only readable in daylight (there were some very simple night signals sent by different light colors, locations and combinations) and then only at close range, by using repeating ships at intervals, signals such as "the enemy is preparing to leave the harbor" could be sent by an inshore patrol vessel to the main fleet fifty or more miles away in a matter of minutes. Signal flags have the disadvantage of flying directly with the wind so that an admiral signaling from the middle of a battle line sailing directly downwind would find that his flags were unreadable by the ships in front of or behind him. To deal with this eventuality, a ship called the repeating frigate would be stationed off to one side or the other of the battle line. From that position the signals from the admiral could be read by the frigate and, when repeated by the frigate, could in short order be read by the rest of the battle line.

In 1837 distant communication, at least on land, was about to change in a way that must have been considered incredible at the time. Samuel F. B. Morse had invented the electric telegraph. The device sent electric impulses by wire. The impulses were usually of two lengths, long or short, or in another version clicks could be used with the interval between them varying from long to short. Codes (Morse Codes) were created to represent letters, numbers and punctuation by combinations of long and short signals ranging from one (E or T) to four for letters, five for numbers, and six for punctuation signs. Eventually one Morse Code was agreed upon for most purposes. The signals were able to be transmitted over any length of wire substantially instantaneously and obviously revolutionized communications overnight. While wire connections were required, the wiring, as well as the sending and receiving stations were quite simple and the system grew from its first link from Baltimore to Washington to cover the eastern United States over the next twenty years. The system was shortly improved by the institution of standard three-letter groups for some usual phrases and other coded short cuts.

It says something about the pressing need for such a device that the telegraph was invented quite soon after electricity began to be understood. The limitations of the telegraph, available only via "hard wire," were relatively minor, for even places not having a telegraph connection could be reached by sending the message to the nearest station and having the message forwarded by messenger. The Western Union messenger, usually on a bicycle, was a common sight in the United States at least for the next hundred years.

The next major advance was the invention of the telephone in 1876, by Alexander Graham Bell, a Scot living in Canada and working in Boston. His invention, which barely beat his competition's to the patent office, converted sound waves in air (the human voice) into electrical impulses for transmission by wire. Now no complicated transmitting and receiving procedures were needed although electricity and continuous wire were still necessary. The sender could speak in a normal tone into a device which would produce the voice signal to be sent to the receiver and answered immediately, permitting a dialog rather than simply sending a coded message and waiting for a reply from the recipient. In addition both parties could speak simultaneously, not always an advantage.

While the telephone was not nearly as much of a revolution in communication as was the telegraph, the telephone had the virtue of being what we would call today user friendly and was an immediate success. One did have to be "hard-wired" to use either system. Transoceanic cables made the wire network international in the course of the century. But still missing was a way to communicate, without wires, to ships at sea or persons moving on land.

Still, the communications gap to be covered by the invention of radio seems slight when compared to what was covered by the telegraph and the telephone. It might appear that, in this regard, the First Decade was of considerably less importance than the years 1837 and 1876 in connection with the communications revolution. In terms of the immediate effect of these inventions that would be correct. Radio was more incrementally useful than revolutionary except at sea where there was no other method of long-range communication. Radio even seemed a step backward in one respect. Bell had freed communicators from the rigors of Morse Code which in practical use had no resemblance to the dots and dashes on signaling flashlights or printed cards. Really good Morse operators can send and receive code at 35 to 50 words per minute. That is the equivalent of 175 to 250 Morse letters per minute. To the uninitiated this sounds like mere noise. The author, who never progressed to anything like this speed,[51] is still in awe of those professional operators, military or civilian, who could not only copy code on typewriters at that speed but could carry on a conversation at the same time. Since the world does not have a great supply of this type of talent, the telephone was a vast improvement for ordinary people.

Enter Guglielmo Marconi. In the 1890s he had experimented with wireless transmission of Morse Code. While credited with inventing radio variously in 1895 or 1897, at that time its range was quite limited, initially about as far as his voice would carry so the practical use was negligible; the same could be said for Nikola Tesla's contemporaneous work in the area. Gradually Marconi increased the range so that in 1901 he successfully transmitted a radio signal from Great Britain to Newfoundland and long distance wireless was born. His system was terribly inefficient in its use of the radio frequency spectrum in that it used spark gap technology. That type of transmission caused most radio transmitters to interfere with each other because they were not operating on specific narrow frequencies. Additionally, the receiving devices of the period were not very selective. Marconi could not have done otherwise for at the time neither the necessary components needed for a narrow frequency signal nor the knowledge to use them existed. But he, by creating a need for different devices, spurred electronic development at a very high rate. The vacuum tube diode, a device useful for rectifying and detecting was invented in 1904 and the triode, an excellent amplifying device, in 1906. Soon radio signals were being generated as continuous waves in fixed and relatively narrow band widths at relatively low power. There was still some interference between signals, but reception was much better. In the early days of

---

51. The author is presently celebrating his fiftieth anniversary as an amateur radio operator with call letters of W5ESU, W3ZXY and K1SXS and license classes from technician to amateur extra.

post spark gap radio, the signals still had to be in Morse Code because the information transmitted came, in effect, from turning the emitted signal off and on. While all of this seems very primitive when viewed from the present, radio was in active use at sea well before World War I because it provided a service not otherwise available.

The rapid evolution of radio continued. First, a method of transmitting sound by imposing audio frequencies on the radio frequencies (amplitude modulation-AM) was developed. It then became possible to transmit voices over the air, making reception much simpler in that one did not have to understand Morse Code although the range of those audio signals was not nearly as great as that of Morse signals. Receiver design was greatly improved, going from the crystal set in the teens (with no amplification and thus requiring the use of headphones to say nothing of other difficulties well known to those who tried to assemble crystal sets as children), to the regenerative (based on a vacuum tube amplifier) with enough power to drive a speaker so the listener did not need headphones and a group could listen to a transmission, to the superregenerative receiver (more of the same) to, finally, the superhetrodyne system, invented in 1913 and used by amateurs as early as the 1920s but not placed into general use until the mid 1930s when some of the essential components had become far more efficient and much cheaper. It had easy and reliable tuning, as much speaker power as needed, good sensitivity for weak signals and great stability. That system is still used today, although when used for AM reception it is limited in the range of its fidelity, is subject to atmospheric interference (lightning particularly) and for public commercial purposes operates on frequencies, in the United States, from about 500 kilohertz to about 1,700 kilohertz. This is in the segment of the electromagnetic spectrum known as medium frequencies, ranging from about 500 kilohertz to about 3,000 kilohertz (or as the number is more usually described, 3 megahertz).

While the high end of the radio frequency spectrum had been identified when x-rays were discovered in the 1890s, the parts of the spectrum below x-rays and above about 10 megahertz, was largely unexplored. A word explanation might be in order here. Frequencies in the radio frequency spectrum are often described in two different ways; either by their wave length in meters or by their frequency in cycles per second. The two are related. The 10 megahertz frequency produces a wave length of 30 meters. The product of frequency times wave length will always equal 300,000 kilometers per second, the speed of light. Thus 30 meters times 10 megacycles (or 10 megahertz, a hertz being one cycle) per second will equal the speed of light. An unusual aspect of radio is its name. All the related concepts, telephone, telegraph and television, contain the term "tele," Greek for "distant." Thus tele-phone—distant sound, telegraph—distant writing, television—distant sight. Radio seems out of place, but really it is not. Radio is simply a short form of the term radiotelegraphy or the transmitting of writing at a distance by radiated signals as opposed to by wire (the name dates from a period when Morse Code was the likely method of imposing intelligence on the radiated signal).

Over time, as a result of experimentation by thousands of persons, both amateur and professional, and the development of new tech-niques and components, the usable area of the spectrum expanded. At the same time transmissions on some frequencies were using tech-niques that reduced the portion of the spectrum needed, particularly single sideband transmission which reduced the necessary bandwidth space for voice although this technique had to wait until the fifties for the necessary hardware to become available so as to make it practicable for general use.

Before World War II there had been created a method of imposing audio frequencies on carrier frequencies, not by amplitude modulation but by shifting the transmitted frequency slightly at audio rates (fre-quency modulation-FM). For technical reasons, this type of transmis-sion requires a large amount of the spectrum for each entity transmitting (a bare minimum of 40 kilohertz for good fidelity trans-mission of music) and actually in the United States 200 kilohertz is allotted to each commercial station so that a full range of frequencies and harmonics can be transmitted without interference to adjacent stations. Military and other non-commercial uses can operate closer together in this mode. In all cases, however, FM is not subject to atmospheric disturbances. By comparison, the AM band commercial station separation is only 10 kilohertz, making for much more efficient use of the spectrum at the cost of atmospheric interference and limited fidelity. By the late thirties, there had developed a competition for operating space in the spectrum which was particularly acute in the high frequency bands (3 to 30 megahertz). The popularity of this area was due largely to the fact that various layers in the atmosphere reflected signals at these frequencies, permitting very long distance communication at low power. The effective use of this feature was, and remains, complex. Different layers reflected or absorbed different frequencies at different times of day, seasons, status of the eleven-year sunspot cycle (really) and different directional paths.

Conferences of most of the world's technologically advanced countries were held in various cities in 1926, 1927, 1932 and 1948. The number of conferences reflects the continuing complexity of uses of the radio spectrum. The purpose was to allocate frequencies for particular uses, sometimes with different uses in different countries if they were sufficiently distant from each other to avoid interference. The result was the structure in use today. Low frequencies were reserved for navigational uses, medium frequencies were largely for AM commercial broadcasts, high frequencies were reserved for various combinations of commercial uses and amateur operations (roughly 3 to 30 megahertz), very high frequencies (30 to 300 megahertz), which were usually line-of-sight, were reserved largely for commercial television, commercial FM broadcasts, taxi and other private communication systems, aircraft communications, navigational equipment and military uses.

As radio evolved with the first half of the twentieth century, various components of the electronics industry also evolved. Sometimes the components were created or improved to fill needs that were felt by the industry; sometimes the components were developed in the abstract and uses were then found for them. Vacuum tubes were perhaps central to this process. From the simple large size diode and triode of the First Decade, vacuum tubes developed in tubes with four, five or more elements, tubes containing, for example two diodes or triodes contained within one glass envelope, complimented by smaller but better resisters, capacitors, inductors, cables, connectors, etc.

Much work was also done on the size and strength of components. Use in mobile operation on land, in aircraft and for various military purposes tended to require special adaptations. Larger tubes and other components were needed for heat dissipation in high power applications. More rugged types of tubes were built to stand the vibration in propeller-driven aircraft or the jarring travel of tanks and, most extraordinarily, to be included in the nose of artillery shells driving miniature radar sets designed to explode the shells at given distances from targets (the variable time (VT) fuse).

Radar (RAdio Detection And Ranging), of course, was an off-shoot of radio as its name readily suggests. It had been known from the 1920s that radio waves could bounce off large metal objects and some of the energy would be reflected back to the transmitter site. Many countries, primarily Great Britain, the United States, Germany and Italy, were working on the concept of radar. Great Britain and the United States worked it out more or less simultaneously in the mid-1930s, with Germany close behind, but the British were far ahead in practical application, having built the first radar net on the Southeast coast of England in 1938, the "chain home" system.

Radar is a good illustration of the impetus war gave the radio and related fields. The British net operated on frequencies of 20 to 30 megahertz. This frequency range was within a part of the radio frequency spectrum that had been used in the 1920s and 30s for long-distance radio work. Thus the biggest challenge initially was not to develop new technology but to adapt the existing technology to handle the large power necessary for radar, and this was quickly done. While radar signals on these frequencies were easily generated, they suffered some serious disadvantages. The wave lengths were so long (10 to 15 meters) that the target definition was not good. For early warning purposes, these frequencies were satisfactory in that one could tell that there were aircraft out there and whether it was a large formation but the precise numbers or even the precise locations were not available. Operating on much higher frequencies was the solution, but devices to generate much higher frequencies and to amplify them to the powerful signals necessary to receive a usable echo from the target did not exist. That is not to say that operation on the lower frequencies did not have some other advantages beyond ease of generation and power of signals. As late as the 1970s the Soviet Union was operating powerful radar systems in the HF range, presumably to obtain over-the-horizon warnings, inaccurate as they might be. As the antenna system swept around, bursts of energy could be heard on high frequency receivers as far away as the United States. This phenomenon was sometimes known in amateur radio circles as the Russian woodpecker.

Apart from improved target definition, higher frequencies permit far better antenna systems. The best size antenna for most radar purposes is approximately one-half of the wavelength of the transmitted signal, Thus, for the early British net the basic antenna elements would be fifteen to twenty-two feet long. This obviously precluded any airborne and many surface uses. Because of efforts to obtain the maximum radiated power from the radar transmitters, multi-element antenna structures would be needed and would have to be in set positions. The towers would have to be stronger because of the wind resistance of the antenna parts, and rotation was then not feasible. Since the early British radar was intended principally to give early warning of Luftwaffe attacks coming from France, the necessity of stationary antennae was not a large problem.

One would have no difficulty imagining how a chain home type antenna being mounted at the height needed and being used for searches with those frequencies would produce only vague evidence of targets but nothing about precise numbers or locations. Thus these sets were obviously useless as fire control devices. It was clear that an increase in the radiated frequency of several orders of magnitude was necessary for radar to realize its full potential. However, the methods for generating higher frequency signals, as for UHF radios for instance, were not very useful at desirable power levels and frequencies for radar. Among other problems, generating signals on extremely high frequencies ran into the issue that the individual waves would reverse polarity between the elements of the generating vacuum tube (there being no significant solid state devices until after World War II), making conventional generating circuits unusable above certain frequencies. While that frequency limit kept rising as smaller and smaller tubes were developed and thus had shorter spaces between elements, the smaller the tube the less its heat-dissipating capacity, thus limiting small tube use severely.

Early in the European war, the British made a vast improvement in radar signal generation when they invented the cavity magnetron, a completely new way of generating extremely high frequency signals at very high power without using conventional vacuum tubes. This may have been the most brilliant invention of World War II, surpassing the atomic bomb for example. Now antenna lengths could be measured in inches, or more usually in centimeters, and efficient turnable antennae could be installed in aircraft, naval and land applications. With that size of antenna and the power generated by the magnetron, antennae could be simplified to a single horizontal element in front of a rotated parabolic reflector, rather like the dish used to receive satellite television. Thus radar could be easily and quickly directed. And the precise target definition afforded by the very high frequencies made it possible to obtain exact positions and numbers of targets. Accurate radar fire control had become possible.

Further improvements followed such that the British and American armed forces had effective radar of all types throughout most of the war. Having effective radar equipment did not end the process of introduction of radar into warfare—the users had to learn how it use it effectively. This is not as easy as it might appear from movies and television programs on the subject but it could be done and it was. The Germans made some progress on radar but not to the same level as the Western allies; the Japanese and the Russians were left well behind.

World War II was sometimes called the "Electronic War" and indeed it seemed to be so at the time. Radio revolutionized battlefield communications then as much as global positioning devices revolutionized navigation and targeting in the Gulf wars. Almost every American unit went into battle with two-way radios linking it to headquarters and to other units. Some of these radios were backpack sized and others were handheld, ideas that would have seemed far into the future just a few years earlier. All were of limited range and limited battery duration as well as being subject to enemy jamming. Notwithstanding the long-range transmission problems which figured so prominently in the Pearl Harbor debacle, long range radio transmissions of orders, intelligence and the like was very common for all combatants. An adverse side effect of all of this radio traffic was the chance that the transmissions would be intercepted by the enemy.

An extreme example of the problem was demonstrated early in World War I by the Russian army which transmitted everything in the clear; all an enemy needed was a knowledge of Russian to read all the traffic. The actual interception of radio signals remains possible but understanding the transmission may be a very different matter. Of course ciphers were often used, particularly for critical or future matters. A cipher is a method of encoding a message by substituting letters and numbers for different letters and numbers according to a plan which both the receiver and sender possess. A code, on the other hand, substitutes one word for another according to a list or matrix which both the sender and the receiver have. The latter is more secure if the number of users involved is small but the former allows for a more widespread use and ease of changing the system if necessary. Ciphers run the risk of being broken by the enemy and the more wireless traffic that can be picked up the better the chances of the cipher being broken.

The other downside of good long-range communications is the temptation it places in front of senior officers to manage matters in detail from afar (micro-management in current parlance). Fortunately the wise commanders resisted the temptation. Admiral Nimitz's orders to Admiral Spruance in mid-1942 are a particularly good example. Admiral Nimitz knew from code-breaking that the Japanese fleet was going to attack Midway Island. It was important to stop the Japanese there before they could menace Hawaii further. The United States Navy was badly outnumbered but the battle had to be fought. Admiral Nimitz gave Admiral Spruance only broad directives and let him fight the battle as he thought best from on the scene even though he was well with radio range of Nimitz's headquarters in Hawaii. The results of the Battle of Midway certainly justified Nimitz's restraint as well as Spruance's judgment.

The importance of radar is difficult to overemphasize although Robert Buderi does so a bit in the title of his excellent book on the development of radar by calling it *"The Invention that Changed the World."* Nonetheless, with radar, modern and reliable radio and sonar all playing major roles, the Electronic War is still a pretty good subtitle for World War II.

The next offshoot of Marconi's invention was television. Technically it predated the war, having been exhibited at the New York World's Fair in 1939 and 1940 but it really did not enter the consciousness of the public until the late forties and early fifties (except for the few of us who appeared, very self-consciously, on the RCA exhibit at the Fair). No discussion is really necessary to describe the evolution or importance of television in the last half of the twentieth century, but it would certainly be possible to debate the value of television. That, however, is beyond the scope of this book and well beyond the competence of the author.

Just the growth of radio, its offshoots radar and sonar and the birth of television might be enough to place radio's descendents as the most important maturities of the First Decade embryos. Notwithstanding this, we have not yet reached the most important descendant of radio, so read on.

Early computers were mechanical in nature and were not particularly reliable. Beginning the late nineteenth century more modern techniques were beginning to be applied to the problem of working out complex mathematical calculations by machines. The first modern computers in the sense we use the word today were probably the Aiken computers invented by Professor Howard Aiken and some engineers from International Business Machines at the beginning of World War II, the first of which was known as the Harvard Mark I. Pieces of it are extant at Harvard University and elsewhere. The Aiken computers would not be recognized as such today. They were huge, room-sized, had no memory, were "programmed" by hard wiring, had large relays at the switch points and had to be fed data in binary form. The end products of the machine were also in binary form. The Aikens did mathematical problems and owed nothing whatsoever to radio.[52] These were the last computers about which one could deny any radio affiliation; their successors owed a great debt to radio. Close behind the Aikens after World War II came ENIAC which used vacuum tubes by the thousands for its switching. It was far faster than the Aikens, much quieter and was sufficiently close to modern computers that some claim it was in fact the first. But note all those vacuum tubes. Had it not been for Marconi or some later inventor with a similar idea to that of Marconi which idea also developed a need for something like a vacuum tube to act as an amplifier or a detector there would not have been

---

52. Some persons (a small minority at least in this country) believe that Konrad Zuse, a computer innovator in Germany during World War II, may have preceded Professor Aiken or at least worked contemporaneously on concepts and machines. They were unknown to each other. For the purposes of this chapter it makes no difference who came first since neither used components derived from radio.

a critical component that could be adapted by a brilliant computer engineer to serve in a capacity not intended or even foreseen by the creator of the tube. In a sense this is almost Darwinian in that an existing element was adapted to perform a function for which it had not been invented. Evolution provides many examples of similar situations. One of the more well known, although somewhat bizarre, was the adapting over time of the jawbones of some reptiles to serve is part of the hearing apparatus of mammals.

It became obvious around the time of World War II that computers were about to become an important factor in some aspects of life. From the Aiken computers doing ballistic calculations for the heavy cannons of the military during that War, the use began to expand. ENIAC was developed shortly after the war and, with what seemed at the time to be an enormous increase in computing power, was capable of taking on far more tasks. A host of additional companies then entered the computer field. As solid state components became a standard part of computers, they became more powerful, smaller and more reliable. But the big problem for most uses was still memory, and particularly random access memory. It was not long ago that a computer with 48K of RAM was state of the art for personal computers (the Apple II, for example) and an upgrade to 64K was available to cutting edge users at extra cost. What this meant was that stored programs of any complexity were not possible in personal computers; the operator had to enter programs for most of what was necessary and even those programs had to be both simple and carefully crafted to use the bare minimum of memory.

The use of vacuum tubes in computers, while short lived, was crucial to their development. But most people today do not remember that era and the concept probably seems terribly primitive. It did work. The author flew a state-of-the-art fighter in the mid-fifties which had a very sophisticated analog fire control computer run entirely by vacuum tubes. When it worked it was quite impressive. Unfortunately it was a troublesome machine since vacuum tubes do not much enjoy 150 degree changes in ambient temperature or high "g" forces. The computer had been designed in 1949 and as we know the transistor had been invented shortly before. There was insufficient time for the transistor to have been perfected, understood and manufactured in bulk for this aircraft but the next generation of all-weather fighters only a few years later had their electronics largely in solid state form, with a dramatic improvement in computer reliability. When solid state technology found a way to pack gigabytes of memory in something that could be easily held in one's hand, the computer revolution really took off. Yet the basic solid state components had not been created for computers; they were created for radio, television and radar, all traceable directly or indirectly to Marconi. Of course, the sequence of development was not always, or even often, from radio to television to computers. Once television and computers became sufficiently viable to be worth spending substantial sums to produce and improve, the developments could well run the other way. Better computer monitor screens produced technology that improved television screens.

A good example of the interrelation of component parts is the cathode ray tube (CRT). A CRT is a vacuum tube invented in fact before the First Decade, usually quite large and shaped like an old-fashioned megaphone. The mouthpiece of the CRT contains the major electrical connections and the wide end is coated with a phosphorescent material. The phosphorescent material glows when struck by electrons. A stream of electrons is generated at the other end of the tube and directed through the tube, being moved in the process by electrostatic plates within the tube or by electromagnetic forces. A CRT is the basic component of an oscilloscope, a device that displays, among other things, wave shapes in circuits.

While radar was being developed, one aspect of the problem was how to display the information received. Contrary to most impressions, the first radar displays were not those of a beam sweeping 360 degrees and outlining the ground or targets. That technology was yet to come although when it did the sweep display was on the face of a

CRT. The first type of presentation was simply a round CRT with a straight horizontal sweep line near the bottom of the tube. When the sweep line was affected by a radar echo, a vertical spike appeared on the line. On the extreme left there was a large spike caused by the radar pulse leaving the transmitter. If a radar echo appeared it showed up as a smaller pulse further to the right. That pulse soon became known as a blip. The distance of the blip from the left spike indicated the distance of the target from the radar set. This type of display thus showed the existence of a target and its distance away. If the radar antenna were stationary, the direction of the target was the direction the antenna was pointing. As radar became more sophisticated with higher frequencies and movable antennae, different types of presentations were developed, but all used some sort of a CRT.

About this time television was being commercialized and the television picture tube was, not surprisingly, a CRT. In this case, however, the CRT beam was deflected in an exceedingly complicated way. There being only black and white television in the early days, when the beam was off the screen remained black; when it was on the screen glowed with a small white spot. Simple enough you say. Well, consider that the electron beam had to trace a very precise pattern of 525 lines on the screen 30 times a second (at least in the United States; some other countries were harder). Nonetheless, it obviously worked. Early computers had CRTs for monitors with operating parameters similar to early television sets. So now we can see how a device greatly improved for use with radio became a critical part of radar and of early computers. But soon radar and especially computers became far more sophisticated and better financed.

A very important aspect of the space exploration efforts of the latter part of century affected radio and related fields quite significantly. Earlier in this chapter the uses of the radio spectrum were discussed and the point was made that the high frequency bands (3 to 30 megahertz, or 100 to 10 meter) were, under the right external conditions, capable of communications over hundreds and even thousands of miles. These frequencies were thus very valuable and coveted by many potential users. This made the allocations of those frequencies important as well as the portion of the spectrum allotted for each user; the narrower the bandwidth used, the more stations which could be crowded into the allocated portion of the band. The situation was made more difficult by the fact that not all parts of even the high frequency spectrum tended to be available for use at any particular time.

Even above 30 megahertz, the competition was acute. In the lower portion of the Very High Frequency band, television channels 2, 3, 4, 5 and 6 (under the United States system) took up most of the area between 54 megahertz and 88 megahertz because each channel required 6 megahertz of bandwidth to transmit both the audio and video portions of the channel. In fact, in almost all portions of the spectrum below about 1,000 megahertz there was severe competition, what with commercial FM stations (starting at 88 megahertz), VHF aviation channels, Ultra High Frequency (UHF) aviation channels, the high end of the VHF television channels (channels 7 through 13), some UHF television channels and a number of other commercial and military allocations.

Unless the radio waves were somehow reflected over the horizon, transmitting stations could only be received on a line of sight basis, or about 50 miles in flat country from a high transmitting antenna. As the frequencies to be used went higher up in the spectrum there tended to be more room available but the chances of having signals that would reflect from various ionospheric layers declined rapidly. Different layers have different reflecting capabilities, depending on the frequency of the signal, the time of day, the time of year and a number of other factors. The situation is further complicated by the fact that if a layer which absorbs a frequency is below one which could reflect it, the signal will never reach the reflecting layer. If there were no usable reflecting or absorbing layer, the signals would pass along the surface of the Earth for a few miles and then would go off into space as the curvature of the Earth dropped the ground surface down relative to the electro-magnetic waves which would be traveling in straight lines.

There was an exception to this rule. In the Arctic area the United States military used to spray non-reflecting frequencies into space at low angles, allowing a small scattering of the signals to bounce back to Earth over the horizon. This expensive procedure (called tropospheric scatter) was used mainly in areas where building intermediate relay stations was not feasible.

Lest it seem that these propagation factors were too complicated to be useful, note that hundreds of thousands of amateur radio operators around the world have used them quite successfully for decades and that monthly projected propagation tables for a variety of frequencies, locations, times and directions are widely published.

But for many years normal long-range radio communications, whether AM or the mid-century development of single sideband, iono-spheric reflection limited as it might be was the only way to go. Enter the space program, and in a few years all of this changed. It became possible to launch satellites that could be so positioned that they were stationary with respect to any point on the Earth's surface at an altitude of about 22,000 miles. Radio signals sent up to the satellites could be rebroadcast downward on line-of-sight frequencies covering, from any one satellite, locations over almost half the globe. Now radio and television signals could be handled by uplink and downlink on frequencies that both were not crowded and were so high that the antennae needed were quite small. The ground receivers could, if necessary, convert the signals back to ones receivable by radios and television sets by encasing the signals in cables so interference over the air was not an issue. Suddenly, at least in geological terms, large portions of the spectrum not previously widely used became usable, the effective range of transmitted signals was expanded enormously and the quality of received signals greatly improved.

If we decide that computers are descendants of radio, albeit through many convolutions, then the test of the most important of the five bits of technology from the First Decade may be whether the computer alone is greater than the automobile, the airplane, atomic reactions or *Dreadnought*.

One short answer might be that the airplane, automobile and nuclear fission are totally dependent on computers today, whereas the computer does not care whether airplanes, automobiles or nuclear reactions exist. The number of computers contained in even a small airplane, such as a seventies military fighter aircraft (designed when computers were just coming into their own) is staggering, They number in the dozens and their importance varies. They can be small and just a convenience, to help to tune a radio, for example, much more serious such as fire control computers, and critical as the computers that handle flight controls in aircraft that are so designed for agility that the aircraft is inherently unstable. If the flight control computers (there is considerable redundancy built into the flight controls for obvious reasons) go out the aircraft cannot be flown manually. Needless to say, *Dreadnought* was dead before the computer was born.

Specific uses are not necessary to conclude that computers are the most important development of the twentieth century. Very few aspects of life in the latter part of the twentieth century were not only touched but dominated by computers. From medicine to document production, from teaching to selling, from sports to publishing, the list is endless. Just taking the last item, publishing, shows the way computer technology has been accelerating. Until the Late Middle Ages in Europe and perhaps somewhat earlier in China written works and copies were made entirely by hand copying. While there were far too few copies made of many potentially invaluable ancient texts, we all owe a great debt of gratitude to those armies of clerks and monks who, through what must have been dreadfully dull work performed under conditions that would not be acceptable for farm animals today, managed to preserve as much as they did for us.

But the invention of movable type and printing presses changed all of that and multiple copies were then far easier to produce. While multiple manually produced copies of documents had been, of course, very helpful in preserving many books and their contents over long periods, particularly from about 500 to about 1500 for the period during which a combination of low literacy rates and political and economic chaos pushed concern for ancient learning far down on the priority lists of almost all in Western Europe, even the extant books had problems. Those small islands in that sea of chaos, generally monasteries, which managed to preserve and copy received learning did not produce exact copies. Inadvertent inaccuracies crept in, compounded by multiple copies with different errors. Other errors, not inadvertent, occurred when semiliterate scribes attempted to correct the received knowledge. The former type of errors could be corrected to some degree by systems developed to find such faults and reconstruct the original texts. For example, a scribe, tired of working in the cold by candlelight, might take a break and, upon returning to work, pick up the text at the word he last wrote. But this might be the same word at a different place in the text being copied. If the scribe had moved back in the text, he would probably recognize this as he wrote the same material again and even if he did not it would be easy for subsequent editors to delete the duplicate passages. Harder to correct were the instances in which the scribe skipped portions of text for the same reason but even these errors could sometimes be repaired by finding another copy of the same text which, although suspicious in other respects, had the missing material. A truer copy could then be assembled. In other areas, scribes copying centuries later sometimes had no feeling for the original language and text and intentionally but erroneously made changes. Some think some of the silent final letters in French appeared because scribes copying French texts, realizing that French was related to Latin, "corrected" the French by putting some Latin inflective endings or Latinized spellings on French words even though they had not been voiced or used for hundreds of years if at all. The "x" representing the plural form of French words ending in "au" may also be the legacy of an unknown scribe who ended the plural forms of "au" words with a flourish which looked like an "x" and so became, over time, a part of the plural form of the word.[53]

In any case, copies made with type tended to be uniform, not necessarily completely accurate but uniform. There was a famous early printing of the Bible that omitted the word "not" in one of the Ten Commandments, a rather significant deviation from the original. Even such a carefully edited and proofread version of the Bible as the King James Version had numerous printers' errors. While found errors could be corrected in one edition, the printing procedures being followed would permit new errors to be introduced and the next edition would tend to contain other errors. Simple typographical errors, while embarrassing in a holy work, would not tend to mislead the faithful but some, such as the introduced "not" in Leviticus 17:14 in the 1611 edition and the deleted "not" in Leviticus 19:10 in the 1613 edition are obviously substantive. In general, however, printing improved accuracy greatly. Still the process of printing was quite slow, both in the setting up of the type and in the printing, limiting the number of books that could be printed. By the end of the nineteenth century the linotype machine made setting type much faster by creating whole lines on blocks of lead-based metal and high speed printing presses, invented somewhat earlier, made newspapers and books more widely and cheaply available. Books still seemed to take longer to produce than was necessary, with the setting of type and proofreading contributing to the delay. By the end of the twentieth century it was possible to set up and print books entirely electronically, with proofreading limited to just the changes made without having to insure that the whole line was correct and had been inserted in the right place. This now unnecessary process was more complicated than it might seem. The result was that computers could do in weeks what used to take months, and most of the time now is waiting for the human input. The printing production process was equally accelerated.

---

53.   Pei, *The Story of Language*, p. 45 (rev. ed. 1965).

We can speculate endlessly about the impact computers will have on the twenty-first century but such speculation has no predictive value. It is almost like natural selection in that we can only take the future one event at a time. While our choices can, in the short run, be logical and beneficial, the long-run accumulation of those choices will undoubted take us in directions we could not dream of today. We can, however, be sure that a hundred years from now the computer function in science, the economy, the arts and everyday living will be greater by orders of magnitude than we can imagine today. Thus these descendents of Marconi's spark gap almost certainly will continue to be the most important of the First Decade's technologies, at least for beneficial purposes.

Conclusion

The Introduction, many pages ago, introduces an analogy between some of the fossils of the Burgess Shale and the technology of the First Decade in that both had five primitive elements which evolved into complex and important areas. I believe the analogy is interesting but it is certainly no more than an analogy.

The time scale of the two categories is vastly different. From the Burgess Shale to the present is about 530,000,000 years, give or take a few percent. The First Decade was about 100 years ago. Thus the Burgess Shale natural evolutionary period is over 5,000,000 times that of the First Decade technology development. This does not mean that it will take half a billion years for that technology to evolve to the same stage that the natural history evolution has presently reached. First of all we do not know, nor could we know, where either of these two evolutions are heading, any more than an animal of the type fossilized in the Burgess Shale could have foreseen the world of the twentieth century or Marconi could have anticipated laptop computer wireless access to the internet. Secondly, natural history is not driven by intent or objective. It happens by random accident followed by an instinctive favoring of any fortuitous result of the accident and change. It is not possible for a reptile lying in the sun in its warm Mesozoic swamp to decide that it would like, for some reason, to be a mammal and realizing that to become one it needed a number of physical changes, including better hearing, for its life on dry land. It cannot just decide to rearrange its jaw bones into a more efficient ear even in the most unlikely event this would occur to the reptile or anyone else for that matter. Although this appears to be the origin of the mammal ear, it did not happen at the behest of an ambitious reptile. Absurd as the idea is, should a reptile somehow succeed in rearranging its jaw bones in this fashion it would avail future reptiles in general nothing because that type of manufactured change cannot be passed down to future generations, Monsieur Lamarck and Comrade Lysenko to the contrary notwithstanding. Parents who have their teeth straightened do not pass straighter teeth on to their offspring except through payments to their children's orthodontist. Natural evolution not only has to wait for some fortuitous random genetic event, it usually needs a series of those events to make any kind of profound change. Further, many of those events may have to happen a number of times before they become imbedded in the species. Consider that an accidental variation toward a more efficient heart might have occurred but the potentially fortunate animal may unfortunately be eaten by something else before it breeds and that variation, useful as it might have been, is lost until it occurs again if indeed it ever does.

In the case of technology, the variations are not random and unintentional. They are directed by human intelligence. This means that they are intended for and are directed toward a preconceived and assumed useful purpose. Obviously this conscious preconceived direction compresses the time scale as does the fact that the variation need happen only once to be carried forward in indefinite numbers. Of course, one variation even if intended of itself may not provide any predictive indication of what the next sequential variation may be. It might be something completely different. Could the inventors of the transistor have anticipated the solid state memory electronics of today? Could the inventors of the differential or the gear shift for automobiles have anticipated the development of all-wheel drive? Nevertheless, the compounding of intentional changes is both infinitely more definite and far more certain of direction than letting nature take its course. Yet in some sense they are still analogous, particularly in their future uncertainty, and therein lies the hope of an interesting future.

But note that, while the automobile evolved, it also begot larger numbers of automobiles and their close relations; trucks, motorcycles, etc. Granted that automobiles became larger, faster and more comfortable. Granted also that their reliability improved greatly, undoubtedly helped to some degree by improved road conditions but helped far more by their becoming a much better product. Most drivers today would have trouble remembering when they had their last flat tire and probably would have difficulty finding the necessary tools in their cars to deal with that type of problem. Seventy-five years ago it would have been an unusual trip that did not involve at least one tire change. Immediately prior to World War II it was expected that the average automobile engine would require a major overhaul by no more than 35,000 or 40,000 miles and would probably have to be replaced well before 100,000 miles.[54] Now we tend to be annoyed if more than oil is necessary in the first 100,000 miles. The automobile bodies seemed regularly to outlast the mechanical parts prior to World War II; the reverse seems the case today. It is important to recognize that, even given the effect the automobile has had on American life in the twentieth century, the technology of the automobile has not radiated much into other areas far removed. Its effect on its environment, while extreme, was narrowly focused.

In the case of aircraft the focus was somewhat broader; in battleships it was very narrow, in radio it was extremely broad and in nuclear reaction it was important but fairly narrow for power plants and potentially catastrophically broad for explosives.

Spindle graphs are a common method of showing the relative prevalence of different types of life present at various times in history. The term comes from the appearance of the graph as a spindle containing thread or yarn to be fed into a weaving process. The thread or yarn may be evenly spread over the spindle, it may be tapered smoothly from end to end, or it may be bunched up in spots. So a spindle graph consists of a basic vertical line usually with symmetrical bulges on either side; the larger the bulge the more types present at that period. The bottom of the spindle is the earliest period covered by the graph,

---

54. Automobile odometers in those days usually would record mileage only up to 99,999 miles.

the top is the latest and in between the waxing and waning of the group is represented.

Thus a spindle graph of algae covering the last two billion years would appear as a slight widening of the vertical line (meaning that there were few species of algae) and the width would be fairly constant (meaning that the number of species did not vary much over time, although there could have been substitutions of one species for another) during the period covered. By contrast, a spindle graph of reptiles would show nothing before about 300,000,000 years ago, expanding greatly in the Mesozoic Era (225,000,000 to 65,000,000 years ago) reflecting in large measure the numbers of species of dinosaurs which evolved in that period. It would contract substantially thereafter in recognition of the extinction of the dinosaurs, unless of course one were inclined to classify birds as dinosaur family members. A spindle graph of angiospermae (flowering plants and trees) would show a first appearance about 120,000,000 years ago and a continual spreading since as the number of species has increased.

This type of graphic display makes no value judgments about the importance of the subject matter or of the individual components. Nor does it indicate the overall number of the individual members. It only shows the degree of diversity among the species represented. The difference can be well illustrated by reference to what a spindle graph of the ape family would show. Beginning some 20,000,000 years ago when apes and monkeys began to go their separate ways, the graph would show an increase in ape species for a while followed by a decline to the present. This seems a sad commentary on the group that contains what we like to think of as the dominant life form on the planet (always ignoring such logical competitors as rats, cockroaches and bacteria of many types), but the graph would show our group's high point (in diversity, at least) as perhaps 10,000,000 years ago.

Having said all of this, were we to spindle graph automobiles, aircraft, atomic reactions, dreadnought-type battleships and radio in the twentieth century, the results would look something like this.

Automobiles-It would have a small bulge ending toward the bottom, representing the decline in diversity for the rest of the century as some of the initial efforts at different types came to naught. Again this does not suggest a lack of importance or a small number of individual units but rather only a lack of diversity. This lack of diversity could be the result simply of having reached an optimal solution but it could also make future adaptations more difficult. This might seem an absurd consideration in the case of automobiles or any other manmade creation. But think how much better off the world would be in the event of a future serious shortage of hydrocarbon fuel for automobiles if hydrogen powered or electric powered cars were available in large enough numbers to have both been perfected in their technologies, reduced their costs and caused to have developed the necessary infrastructures to support a large number of such machines. The issue is quite parallel to that of natural history.

Aircraft-Again there would a small bulge in the First Decade and a few years thereafter (showing the lighter-than-air variations). This bulge would have tapered off but another bulge would begin in the thirties with the short-lived auto-gyro followed shortly by the helicopter, jet and rocket engines, hovering aircraft and other distinct types, producing a slow but steady increase in width for the last seventy years.

Atomic Reactions-The graph would have its first expansion halfway up the graph in 1945 when the first bombs were exploded. There would be a further widening in the 1950s to reflect both the development of the hydrogen bomb and the atomic power plants of several varieties, a width it maintained until the end of the century.

Battleships-It would be fair to say that the battleships and battle-cruisers which followed *Dreadnought* were all of the same species, as always using the word loosely since so far as is known ships do not breed and produce fertile offspring (the definition of a separate species). Of course there were significant differences of various sizes and numbers. The sizes ranged from 20,000 to 70,000 tons, the armament from 12" to 18" main batteries (a hitting power increase of perhaps a bit more than three times) but essentially the species was the same. After all, the species *canis familiaris* covers dogs from toys to mastiffs, probably a greater difference than appears among battleships. The lack of diversity in the battleship group neither contributed to its early demise nor detracted from its importance during its brief life. The fact that the spindle would become bare at mid-century is irrelevant; the effect of *Dreadnought* would have been felt well before then and that effect would continue to plague the world.

Radio-Its spindle would gradually enlarge from the beginning with the development of the continuous wave, followed by amplitude modulation. Further expansion in the thirties would come from frequency modulation, the beginnings of radar and the development of television. When computers were thrown into the mix the expansion would become greater and it is still expanding today although the rate would depend upon the number of "species" contained within the overall category of "computers."

Another very important, perhaps the most important, difference between the First Decade developments and those indicated by the Burgess Shale has to do with "what if."

What if Henry Ford had decided to go into another line of work after one of his early automobile failures? Well, the automobile had been developed before the creation of the Model T. Without it automobile development would have continued. In all probability good machines would have been built and improved rapidly. Missing would have been perhaps some of Ford's engineering innovations, for better or worse, and his production techniques with the related cost savings, clearly for the worse. The automobile would not have had nearly the sales it did in the teens and twenties without Ford. As a result, the pressure to improve roads and services would not have been as acute as it was. Eventually, however, this would all have happened and in the long run the automobile would have secured the place it did in American society. Thus, only the timing seems likely to have been affected by Henry Ford's absence, although by how much is impossible to say.

What if the Wright brothers had stuck to the bicycle business in Dayton? Clearly heavier-than-air powered craft would have been delayed. But many good minds were working on the problem and the concept of lift had been amply displayed by the success, or semi-success, of gliders. Remaining major issues were a satisfactory power supply and effective controls, particularly of a coordinated type. It is almost certainly the case that in the next ten or fifteen years someone else would have found a solution for both of the problems and in all probability the solutions would have been more elegant than those of the Wright brothers. In that context a look at a picture or model of either the engine or the controls is enough to convince any modern pilot how lucky the Wrights were that their first aircraft worked as well as it did. The Wrights were most fortunate in succeeding since not only did they have a shaky engine and dubious controls, they had no experience with a powered aircraft nor had they any experience with control pressures. If one is to build an elevator to change the pitch of an aircraft, the present state aeronautical engineering would allow the designer to calculate the appropriate size and range of movement of the elevator under a variety of ambient conditions. But in 1903 this would have had to be done by guesswork, for data from gliders would not serve very well. Too little area and/or too little movement would mean that pitch control would not work leading to an inability to correct for wind gusts or other events causing inadvertent climbs or dives; too much area and/or too much movement would cause over-controlling, a serious problem as well. unable to be corrected by the pilot. The former would be likely to produce a stall, followed by a spin; the latter would cause excessive movement of the aircraft, a serious problem as well, probably also leading to a crash. The Wrights ran a grave risk, on this factor alone, of having their first flight being their last. Had the Wrights stayed on the ground or been killed testing their machine, heavier-than-air craft would have been set back a few years but if alternative heavier-than-air machine had been built before World War I, the war might well have caused enough accelerated improvement to make up the lost time. Long run, perhaps little would have been different.

What if the United States had decided not to build the atomic bomb or, having decided to build it, was unsuccessful in its efforts? The two other nations most likely to have been involved in a similar attempt were Germany and the Soviet Union. Germany had available the basic atomic science although it was missing a number of important central Europeans with the potential for great contributions to that effort—Einstein, Bohr and Teller, for example. Germany's objective would have been to build the bomb regardless of the United States activity because the bomb could have been a war-winning weapon for Germany. But it gave up on the effort early on, perhaps because of a shortage of the requisite resources or because of technical dead ends. The Soviet Union, on the other hand, while lacking in such resources as well, did not need the bomb for World War II; it became interested only as a competitive matter when the United States succeeded. The potential resource shortages might have been overplayed with respect to Germany and the Soviet Union in that neither country needed to devote the resources to the effort that the United States did. The United States was determined to build the bomb and end the war as soon as possible. As such, resources were thrown at the project at a rate only the Americans could afford. The result was a serious case of what we might call over-engineering. For example, since uranium was a raw material of choice and only the U235 isotope of uranium was usable in a bomb, somehow that isotope had to be separated from the far more dominant isotope, U238. Because they were isotopes of the same element, no chemical reaction could differentiate between them. The three theoretical choices for separation were difficult and expensive; first, forcing a gaseous uranium compound (in spite of the heaviness of the uranium atom, there was at least one, UF_6—uranium hexafloride) through a series of semipermeable membranes during which the molecules containing the U235 would move slightly faster, increasing their concentration at each stage until the U235 content was high enough;

second, centrifuging the element forcing the slightly heavier U238 to the outside; and third, converting the uranium to a plasma which when forced through an electric field would cause the similarly charged but lighter U235 to deviate more. All of these methods could work but the gaseous diffusion method was by far the most efficient. The United States tried two of the three; a competitor needed to have built only one, at least if it were gaseous diffusion. Additionally, a competitor could have gone with either uranium or plutonium as the active element in an atomic bomb; the United States made one operational bomb of each.

If, however, the United States had not built the bomb and if Germany had dropped the project and if the Soviet Union had not started, the biggest change would have been the necessity for the United States to invade the Japanese home islands for it does seem certain that but for the atomic bombs the Japanese would never have surrendered. The casualties on both sides would have been enormous and the additional destruction to Japan might have made it impossible to reestablish the country. In addition to preventing casualties, the economic cost saving to the United States by building the bomb and its delivery system as opposed to the cost of the invasion was quite substantial. Without the bomb the Cold War might possibly have been less tense, but the reverse might also have been the case because the "mutual destruction scenario" would not have been credible without atomic weapons. That scenario seems to have been the principal reason the Cold War did not turn "hot" at some point between the late 1940s and the early 1990s.

What if Admiral Fisher had thought that submarines were the ships of the future? He would have been many years ahead of history but *Dreadnought* would not have been built. Unlike the situation with automobiles, aircraft and radio, we do not have to speculate or to assume that, of many others working in the field, someone or some group would complete the inventive or development process so that a ship would have been completed similar to the historical ship. We know in considerable detail what the alternative dreadnought would have been like and when she would have been completed. Not only had the concept been published in 1903 by the famed Italian naval architect Vittorio Cuniberti in *Jane's Fighting Ships,* probably the most respected public source for matters naval, but also the United States Congress had authorized two dreadnoughts[55] in 1904 and had appropriated funds in 1905, although construction was not begun until 1906 nor completed until 1910. While the American ships varied from *Dreadnought* in type of propulsion (the Royal Navy's was better) and armament position (the American's was better) there was no question but that the American ships would have had the same effect as *Dreadnought,* to make obsolete all other battleships in the world. As indicated by the delayed completion dates of the American ships, the only difference would have been a delay in the start of the naval arms race between Great Britain and Germany by at least half a dozen years. While this does not seem much, if the assassination of Archduke Ferdinand had still happened in 1914, the arms race would hardly have begun in terms of ships completed and commissioned. Germany would have had no chance at sea except via U-boats which in 1914 were an unknown quantity notwithstanding the outstanding success of *U9* in sinking three Royal Navy cruisers in an hour in 1914.[56] Thus, would there have been World War I, the Russian revolution, World War II and the Cold War? Who knows for certain, but the odds are "no." Obviously other serious events would have occurred in the twentieth century to respond to the geopolitical strains of the nineteenth and early twentieth centuries but it is hard to imagine that they could have been as serious as the actual ones.

What if Marconi had been electrocuted in the course of his early experiments which, given the power levels he was using, was entirely

55. *Michigan* and *South Carolina* were, of course, not called dreadnoughts at the time because the term was not in use until 1906 but their design was consistent with that of *Dreadnought.*

56. The commander of *U9* in the one-hour attack on HMS *Cressy, Hogue* and *Aboukir* was subsequently transferred to command of *U29* and was killed when that submarine was rammed by *Dreadnought.* See p. 72.

possible? There is a good probability that Tesla would have invented radio in short order; he was part way there anyway. If he had, presumably the rest of the radio-radar-computer sequence would have evolved pretty much as they did with perhaps two important exceptions. If the timing of radio-radar had been delayed a few years radar might not have been at the level it was in the late thirties. That would have had two important consequences. First, the British would not have had a radar net covering its southeast coast in time to be of critical importance in the Battle of Britain (assuming the timing of World War II stayed the same). If, as a result, the Battle of Britain had been lost by the British, Britain might well have been invaded and conquered in the late summer of 1940, with disastrous consequences for the allies.[57] Second, the pressure would not have been on to develop a high-powered, high frequency radar signal generator. Since that development (the cavity magnetron) was not the result of a particular development program but was a stroke of such genius that, even when confronted with both the hardware and a detailed schematic diagram, only one of the top physicists in the United States could figure out how it worked. As an aside, fifteen years later the author's quite sophisticated electronics course in the United States Air Force did not, or could not, explain the operation of the cavity magnetron. It was only when the author ran across a report of Professor I. I. Rabi's explanation to the assembled scientific brain trust that "it worked like a whistle" that the concept made sense, at least to the author.[58] That invention was such an unusual stroke of genius and was made under such pressure that it might well not have happened at any other time.

The other event that might possibly have been a one-time occurrence was the invention of the transistor, starting the solid state revolu-

---

57. Some historians feel, for what may be good reasons, that a German invasion of England would have failed in any case. *E.g.*, Murray, *German Military Effectiveness*, p. 165 (1992); Price, *Battle of Britain Day*, p. 130 (2000).
58. Buderi, *The Invention that Changed the World*, p. 49 (1996).

tion. Whether that was inevitable under other timing circumstances is at least arguable.

This brings us to the real difference between the evolution of the First Decade technology and the natural evolution of the Earth, True, the time scale is vastly different; true we have control to a large degree over the technological evolution and true had some parts of the technological evolution not gone as they did our lives would have been different; in the case of *Dreadnought* our lives could have been, and in the case of atomic fission and fusion might still be, dramatically different. But what if any one of the millions of random natural evolutionary events had gone differently? Even if nature somehow had the opportunity to try to rerun the evolutionary events of the last half billion years over again, the chance of millions of those random events coming up with the same results is effectively zero, producing results that could only be very different, possibly better, possibly worse, but we in anything like our present form would certainly not be here to see them![59]

By random in this sense what is meant is that changes occurred spontaneously in a wide variety of forms. To the extent that the change was, in some way, detrimental to the life style of the animal receiving the change, that animal might have had a lesser chance of surviving to breed and thus to pass on the newly-acquired trait to its descendents. To the extent that the change was helpful to the life style, it enhanced the chances of the animal surviving to breed and thus perhaps passing that trait on to the advantage of future generations. None of this is certain; it is only a matter of probabilities. Small changes and low probabilities made for very slow evolution. As it does not follow that advantageous changes will always produce good future results, it does not follow that genetic diseases or conditions that indicate a propensity in individual animals (including humans) toward a disease are traits that will necessarily cause that disease to be eliminated by virtue of the disadvantage of the condition. It would seem logical that, over time, those disadvantaged individual animals should be eliminated, although

---

59.   Gould, *Wonderful Life*, pp. 318-23 (1989).

if the disease susceptibility were to conditions that occurred after breeding age there should be no effect on the species. If the disease had killed off, or made sterile, the affected animals, many diseases should be self-eliminating due to natural selection. We know that at least currently this not so but that is not to say that some diseases were not self-eliminated in the past.

Suppose that all of those million plus changes necessary to cause the evolution of the creatures of the Burgess lagoon into us were not in fact random. If they were like coin flips, then the statistical chances of the same sequence appearing over a million flips is clearly zero. But if there were other factors affecting the evolution so that the changes were not entirely random, the odds might be at least somewhat better for a second run to resemble the first.

Two possibilities come to mind. One is some variety of divine direction. Almost by definition this is something that cannot be discussed logically. If, however, the divine purpose had been to create human beings, the procedure followed was extraordinarily slow and full of missteps. This hardly seems like the course some sort of controller of our destiny would take but the acceptance or rejection of this possibility is really a matter of religion, not science or logic.

The second possibility might assume that, while there are still a million or so choices, they might not all, or any of them, be random. There might have been some type of predilection for some reason we cannot discern along the path of evolution pushing the course of evolution, gently perhaps, along the path it actually followed. Were that the case even if that predilection could not be determined it would make the chances of a repeating of the same evolutionary path somewhat more probable although still highly unlikely. Any such predilection might somehow influence gradual evolution but could hardly deal with two major species-destroying events.

At the end of the Permian Period, about 225 million years ago, and again at the end of the Cretaceous Period, about 65 million years ago, events occurred which caused the extinction of a great many species in a short (by geologic standards) time. The Cretaceous catastrophe was probably caused by a large meteor impact near what is now Mexico. The impact presumably threw up so much debris into the atmosphere that the amount of sunlight reaching the Earth's surface was substantially reduced for a period of years, producing dramatic climatic changes. While this might seem at first glance unlikely, in recent years volcanic eruptions (particularly Krakatoa in the nineteenth century) have caused world-wide effects from their comparatively small dust distributions into the atmosphere. The species lost from the meteor strike were very significant, including all species of dinosaurs which had survived to that date. The result of the dinosaur destruction was to make room for a huge radiation of mammal species. Mammals had been around since before the rise of the dinosaurs, but the mammals were kept to small size by the domination of the land first by large reptiles and shortly thereafter by large dinosaurs. When the dinosaurs died off, mammals, perhaps because of their small size or perhaps being warm-blooded animals, could survive the colder temperatures better than the cold-blooded dinosaurs and could in either case move easily into the niches for larger animals.

The great dying of the Permian Period also could well have been caused by a meteor strike although the evidence is not nearly as clear as it is for the Cretaceous, and the fact that the Permian event affected sea life so much might suggest another cause. The Permian disaster had another twist to it in that, while perhaps 95% of all of the then existing marine species became extinct, no whole type of creatures was without representation after the great dying was over. It was thus easier for a full range new marine species to evolve from the surviving remnants of the Permian sea life. In the Cretaceous, since all of the dinosaurs disappeared, the mammals were spared the difficulty, perhaps the impossibility, of competing. They just took over all levels of the dinosaur territory by default. A similar situation developed with respect to placental versus marsupial mammals. While both evolved similarly and marsupials may have come first, eventually when the placentals had appeared and the two groups were in direct competition, the placentals usually prevailed. In North America today the only common marsupial is the opossum, whereas in Australia, New Guinea and Tasmania which separated from Asia before the placental mammals had appeared at least in that area, marsupials developed which filled categories occupied by placentals elsewhere.[60]

Recall that on page x *Pikaia gracilens* was introduced as the first of the chordates. It is, of course, possible that other chordates had been evolved at the time but they have not been found. Thus a two inch, soft bodied, boneless marine animal may have been the ancestor of, among other things, all vertebrates, including us. We, being homocentric or at least vertebratecentric, should have to view the possible extinction of *Pikaia* and its progeny during one of the great dyings or otherwise as great a disaster as could befall the Earth but of course we would never have known of it.

With all of this confusion, turmoil and extinction it is difficult to believe that any predilection could have had much of an effect on the evolutionary progress. We are thus left to ponder the extraordinary benefits of the rapid and well orchestrated technological evolution of the past century in our material lives and the excruciatingly slow, haphazard and imperfect evolution of our species. We, as a group, can take great pride in the results of the first and can only thank fortune for the results of the second. But, on a more somber note, as a result of one of the First Decade technologies we have in our hands the ability to bring down upon the Earth the equivalent of the great dyings of the Permian or Cretaceous Periods. It is greatly to be hoped that we shall be able to handle the potential downside of our technology as well as we have handled the upside, not perfectly but adequately, so far.

---

60.   The third group of mammals, the monotremata, have been left out of the comparison because, with only three species (one platypus, two anteaters) and a limited range (Australia and New Guinea) they are not a big factor, although a very interesting one.

0-595-66193-9

Printed in the United States
17779LVS00001B/73-84